AutoCAD 2020 from Zero to Hero

AutoCAD 2020 from Zero to Hero

Zico Pratama Putra
Ali Akbar

Kanzul Ilmi Press

2020

First Printing: 2020

ISBN-13: 9781080727902

Kanzul Ilmi Press
Woodside Ave.
London, UK

Bookstores and wholesalers: Please contact Kanzul Ilmi Press email

zico.pratama@gmail.com.

Trademark Acknowledgments

All terms mentioned in this book that are known to be trademarks or service marks have been appropriately capitalized. AutoCAD, Inc., cannot attest to the accuracy of this information. Use of a term in this book should not be regarded as affecting the validity of any trademark or service mark.

AutoCAD is registered trademark of Autodesk, Inc.

Unless otherwise indicated herein, any the third-party trademarks that may appear in this work are the property of their respective owners and any references to the third-party trademark, logos or other trade dress are for demonstrative or descriptive purposes only

Ordering Information: Special discounts are available on quantity purchases by corporations, associations, educators, and others. For details, contact the publisher at the above-listed address.

Contents

Chapter 1 Introduction to AutoCAD 1

  1.1 What's New in AutoCAD 2020? 1

    The Look...1

    Measure Tool...2

    Block Tool Palette..2

    Purge Command Improvement...3

    DWG Compare...5

    External Storage Providers 5

    It Doesn't Seem Much..6

  1.2 A Glimpse of AutoCAD 2019? 7

  1.3 Creating an AutoDesk account 11

  1.4 Install the software 16

Chapter 2 Get Around in AutoCAD 20

  2.1   Why AutoCAD? 21

  2.2   XY Coordinate...22

  2.3   Angle in AutoCAD 24

  2.4   Inserting Point in AutoCAD...25

  2.5   AutoCAD's User Interfaces 26

    2.5.1 Change Units in AutoCAD..27

    2.5.2 Explanation of the Workspace..27

    2.5.2   Ribbon...29

    2.5.2   Menus..32

  2.6   Open Drawing...39

  2.7   Close Drawing...41

  2.8   Export as PDF 42

Chapter 3 Drawing in 2D 45

  3.1 Set up Snapping 45

  3.2   Create 2D Drawing..46

    3.2.1   Drawing a line 46

    3.2.2   Drawing Polyline..53

    3.2.3   Drawing a circle...62

    3.2.4   Drawing The arc 68

3.2.5   Drawing Rectangle 77
3.2.6   Drawing Polygon..83
3.2.7   Drawing Ellipse..86
3.2.8   Drawing Hatch 89
3.2.9   Drawing Spline...93
3.2.10   Drawing XLINE..96
3.2.11   Drawing RAY...97
3.2.12   Divide 98
3.2.13   Drawing Helix 100
3.2.14   Drawing Donut 103
3.3   Modify 2D Drawing 104
3.3.1   Move..104
3.3.2   Rotate..106
3.3.3   Trim...109
3.3.4   Extend 115
3.3.5   Erase..117
3.3.6   Copy 118
3.3.7   Mirror 120
3.3.8   Fillet 122
3.3.9   Chamfer 124
3.3.10   Explode 125
3.3.11   Stretch 127
3.3.12   Scale 130
3.3.13   Array Rect...132

Chapter 4 Case Studies 134
4.1   Create Simple House Plan.......................................134
4.2   Create Simple Gear...148
4.3   Create Simple Piston 156

Chapter 5 Draw 3D Drawing..164
5.1   Configure 3D Workspace 164
5.2   Draw 3D Objects 165
5.3.2   Draw Box...165
5.3.3   Draw Cylinder...171
5.2.3   Draw Cone 174
5.2.4   Draw Ball 177
5.2.5   Draw Pyramid...179

5.2.6   Draw 3D Donut..180
5.2.7   Extrude 2D Object...183
5.2.7   Chamfer and Fillet Feature 185
5.2.8 Merge, Subtract, and Intersect 3D Objects.........187

Chapter 6 Mesh-Files in AutoCAD **189**
6.1 Import .stl and other Mesh-Files.........................**189**
6.2 Export .stl **190**

Chapter 7 Create Technical Drawing................................**191**
7.1 Insert Model Views **192**
7.2 Place Dimensions **194**
7.3 Detail and Section View...**195**

About the Author...**198**

Can I Ask a Favour?..**199**

CHAPTER 1 INTRODUCTION TO AUTOCAD

Welcome to the AutoCAD's World. This AutoCAD tutorial will teach you the basics of using AutoCAD and create your first objects. AutoCAD is a robust tool for creating 2D and 3D objects, like architectural plans and constructions or engineering projects. It can also generate files for 3D printing. If you want to start this AutoCAD tutorial for beginners, you should have about an hour to do so.

1.1 What's New in AutoCAD 2020?

During the first quarter, Autodesk released its new software as usual. AutoCAD first. People are used to being excited to see what's new, I remember. But things have changed. AutoCAD hasn't changed much in many years.

Integrating the toolsets into AutoCAD 2019 was a big step, but they're not new – as I will describe later in this section. The last version I remember has major improvements with AutoCAD 2008. If you're working with 3D, you could say 2010.

AutoCAD 2020 has nothing different. So, what's new in AutoCAD 2020?

The Look

The Interface is updated, of course. We know that AutoCAD always has an "updated" look. It looks sharper and easier to look.

This version is claimed faster than before. Even the installation is faster. This improvement is important. I don't expect Autodesk will change AutoCAD much, but they obviously need to improve the performance and reliability. Faster and less crash, everyone will be happy.

Measure Tool

You can use the live measure tool just by moving the cursor. AutoCAD will show the live measurement between the nearest geometry according to your cursor position. The Measure dropdown list on the Utilities panel has been a handy way to find distances, angles, diameters, and areas for years. Traditionally, the user selects the desired measurement option, selects the relevant objects (or starts at points) in the drawing, and notes the measurement. A faster version of this tool is now available, called Quick Measure.

The program finds measurements near the crosshairs and automatically displays them when this tool is active. It keeps track of distances along line segments, distances between parallel lines and angles (with the usual right angle icon). Quick measurement is now the default measurement tool that you can access with the MEAsure command. The previous subcommands are still available.

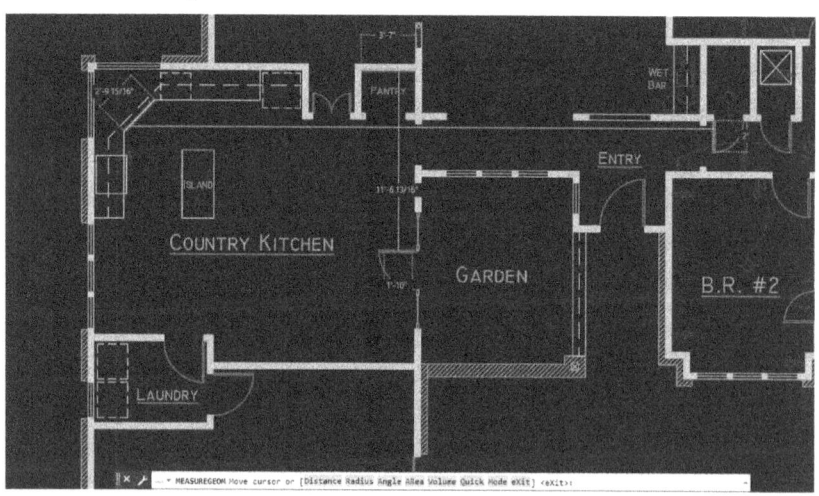

Pic 1.1 Image source:
https://blogs.autodesk.com/autocad/introducing-autocad-2020-autocad-lt-2020/

Block Tool Palette

I like the Tool Palette, but the Block Palette can be better. Autodesk changed many dialogs to Palettes. Many of them are pretty useful. I expected that by using Block Palette, we could change the scale, rotation on the fly when placing a block. But it works pretty much like the old dialog box. The difference is now you can access the block without closing the dialog box.

It looks more compact and easier to use, but it doesn't offer any new functionality. Maybe in AutoCAD 2021, we can do more in the Block Palette?

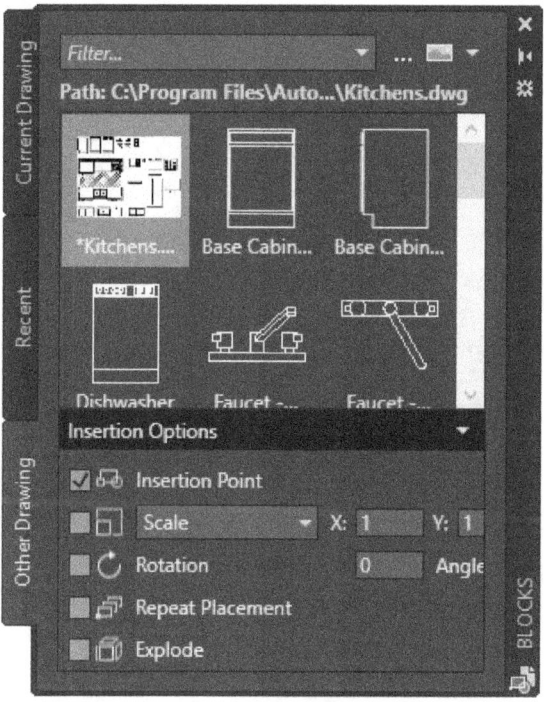

Pic 1.2 Block Palette

And of course, if you prefer to use the old dialog box, you can use CLASSICINSERT command.

Purge Command Improvement

The purge dialog box is updated. Autodesk changed the terminology from "View items you can purge" to "Purgeable items", but they are exactly the same.

It offers the same functionality, but now you can see the preview pane on the right side.

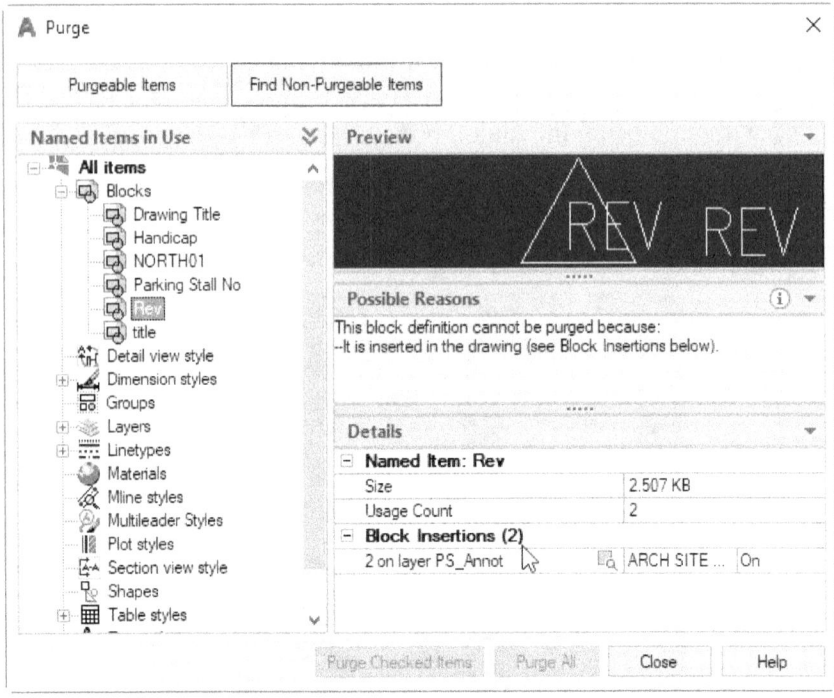

Pic 1.3 Purgeable items

There is one thing that is pretty handy: It helps you to find the object and gives better description on non-purgeable items. It tells you how many items on the drawing, on which layer, and it even tells you in which block it is for a nested block.

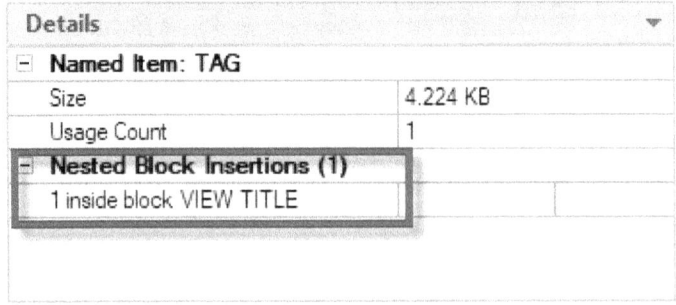

Pic 1.4 Non-Purgeable items

DWG Compare

Autodesk introduced DWG compare on AutoCAD 2019 to see the differences between two drawings. It was useful but there is one annoying thing about it: it compares the drawings on a temporary drawing tab. It means you can't annotate the difference or give a markup.

In AutoCAD 2020, AutoCAD compare the drawing within the file. It allows you to draw a revision cloud or other annotations. The COMPAREEXPORT command exports a similar "snapshot" drawing, replicating the 2019 method.

The COMPAREIMPORT command (or button) will bring selected, unique objects from the comparison drawing into the current drawing. The checkmark dismisses the editing state in order to return to business as usual.

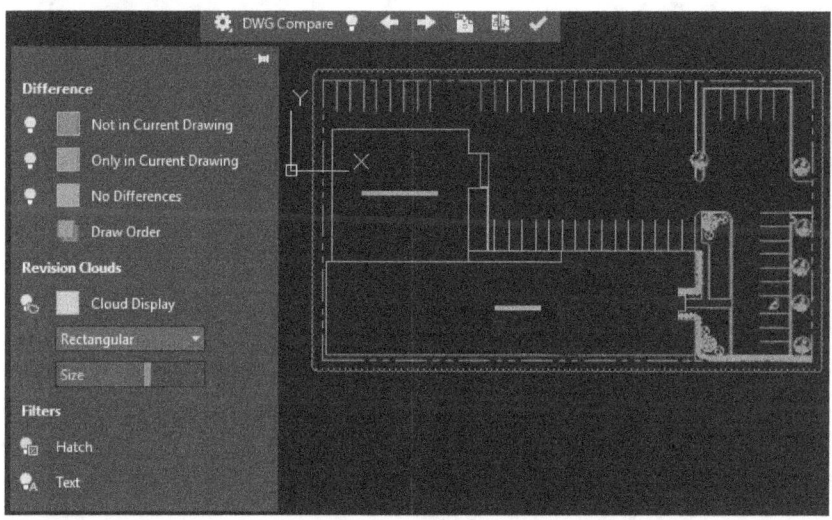

Pic 1.5 DWG Compare

External Storage Providers

Autodesk claims that AutoCAD 2020 works well with other cloud storage providers. I was confused. Any program can work well with any cloud storage providers, and I have used Dropbox and OneDrive for years. So why this is new?

It turns out that this new feature is for AutoCAD web, not the desktop. Now you can use another popular storage provider to work with AutoCAD web.

It is definitely useful if you use AutoCAD web and save your files on those online storage.

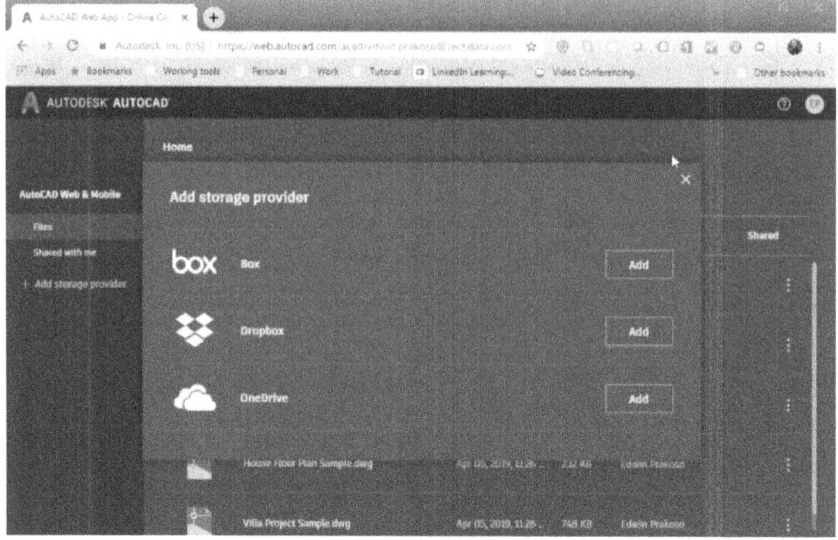

Pic 1.6 External Storage

It Doesn't Seem Much…

Honestly, it doesn't look much. Revit 2020 has more new features than AutoCAD 2020. Maybe because AutoCAD is a mature product as a drafting tool. Maybe because of AutoCAD users consistently refuse to use new features and keep trying to make AutoCAD "normal" like the previous version. Not many users use Dynamic Block and Sheet Set, for example.

Maybe Autodesk focuses more on work under the hood, make it faster and more reliable.

Have you tried AutoCAD 2020? How do you like it?

1.2 A Glimpse of AutoCAD 2019?

Let's begin with an essential look at the previous AutoCAD 2019. The three key parts of this release are the desktop application you're likely familiar with, AutoCAD Web, and finally what Autodesk is calling One AutoCAD. While each of these elements brings positive changes to AutoCAD, the introduction of **One AutoCAD** is likely the most impactful.

One AutoCAD represents the most significant change to the way Autodesk packages AutoCAD in recent memory. Starting today, AutoCAD and its many vertical flavors are no longer offered as individual subscriptions. Instead, AutoCAD and AutoCAD-based products such as AutoCAD Architecture and AutoCAD Map 3D are combined into One AutoCAD. This combined product includes access to AutoCAD plus the entire portfolio of AutoCAD verticals which are now called Toolsets.

The Toolsets included in One AutoCAD are:
- Architecture (previously known as AutoCAD Architecture)
- Mechanical (previously known as AutoCAD Mechanical)
- Electrical (previously known as AutoCAD Electrical)
- Map 3D (previously known as AutoCAD Map 3D)
- MEP (previously known as AutoCAD MEP)
- Raster Design (previously known as AutoCAD Raster Design)
- Plant 3D (previously known as AutoCAD Plant 3D).

Although all new subscriptions to AutoCAD include access to the specialized toolsets outlined above, existing customers with eligible subscriptions may opt into a One AutoCAD subscription for the remainder of their contract. Additionally, while AutoCAD subscribers must opt-into the new One AutoCAD, it will be included as part of a forthcoming update to the Architecture, Engineering, and Construction Industry Collection.

Note on Civil 3D: Even though AutoCAD Civil 3D is an AutoCAD-based product, it is excluded from the One AutoCAD package, and will be renamed Autodesk Civil 3D upon its next release.

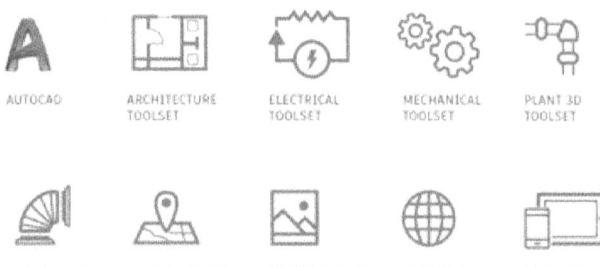

AUTOCAD   ARCHITECTURE   ELECTRICAL   MECHANICAL   PLANT 3D
           TOOLSET   TOOLSET   TOOLSET   TOOLSET

MEP TOOLSET   MAP 3D TOOLSET   RASTER DESIGN   AUTOCAD   AUTOCAD
                                 TOOLSET   WEB APP   MOBILE APP

Pic 1.7 One AutoCAD includes a variety of vertical extensions

The Drawing Compare tool. This is like the AutoCAD Architecture compare feature, but with a few additional abilities to cycle through areas that have changed on both drawings and xrefs. It's not a new feature so much as a port of the existing DWG Compare tool, with a useful revision cloud feature to highlight changes for added visual confirmation of changed areas.

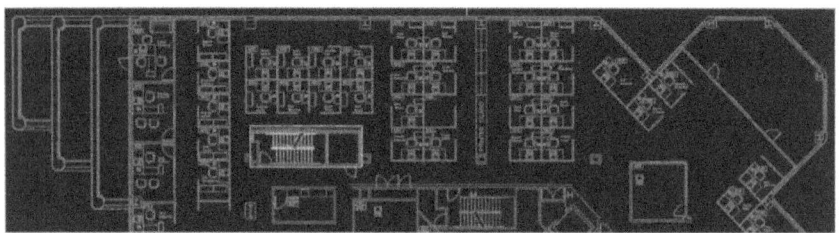

Pic 1.8 The green and red graphics highlight the differences between the first version of the drawing (green) and the second version (red)

Enhancements to Share Design Views. These are intended for those AutoCAD users who wish to share views of drawings via a web-based interface without actually sending DWG or PDF files. Triggered from AutoCAD, Shared Views are then sent as a link to anyone, and can be viewed and commented on via a web browser without any special application software to install. Any comments stored in the shared view can then be brought back into the AutoCAD application by the original author.

AutoCAD Web and Mobile apps. While the new features inside AutoCAD 2019 expand and refine existing workflows, AutoCAD Web seems to open the door for entirely new workflows. Autodesk talked a lot about the modernization of the AutoCAD codebase during the launch of AutoCAD 2018 last year. Rather than an enhancement to AutoCAD for your desktop, these are adjunct tools that facilitate DWG editing on web browsers and mobile devices.

AutoCAD Web is not just a continuation of the former AutoCAD WS and AutoCAD 360 you may have dabbled with in the past. Instead, AutoCAD Web is a desktop-class experience delivered through a web browser. Autodesk achieves this by powering AutoCAD Web with the same engine as the desktop version of AutoCAD 2019. Because of the code modernization efforts discussed last year, we now have a web-based version of AutoCAD that matches the power and performance typically reserved for desktop applications. Testing this functionality for myself, I was genuinely impressed at the ability to draft a simple architectural floor plan from scratch using AutoCAD Web. What impressed me even more was the performance of that experience easily matched the overall performance of AutoCAD 2019.

Though I don't see full-time users of AutoCAD trading in their subscriptions for AutoCAD Web just yet, I do see it serving as an incredible supplement to many workflows that take users out away from their desktop. Likewise, I see AutoCAD Web as a possible alternative for passive users of AutoCAD. People who need to perform basic edits to drawings, but who do not spend a majority of their day using AutoCAD. Since AutoCAD Web isn't tied to the typical release cycle of the desktop version, I am very eager to see what Autodesk adds to the web experience throughout the coming year.

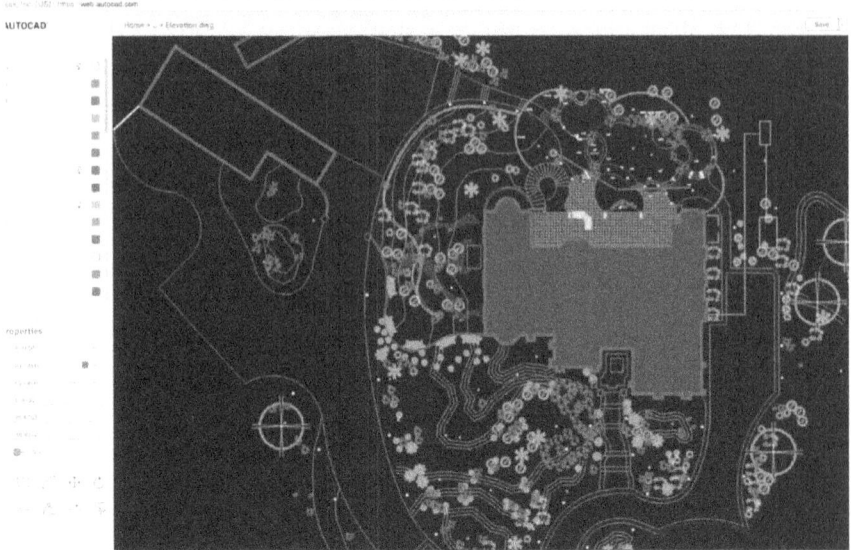

Pic 1.9 AutoCAD 2019 introduces tools for DWG editing on web browsers and mobile devices. This screenshot depicts manipulating AutoCAD files in the web.autocad.com portal

2D graphics performance updates. Functions usually requiring redraw or regen such as draw order, zooming, panning, layer properties, or displaying raster/PDF overlays are reported to work two times faster.

Shared Views. With just a click, you can now share drawing views with anyone thanks to a new integration with the Autodesk Viewer. The Autodesk Viewer is a web-based tool that lets anyone view a vast array of Autodesk file formats, including DWGs, without installing anything. Shared Views offer a refreshing alternative to ordinary collaboration workflows that require teams to convert their drawings to PDF, email those PDFs to stakeholders, and finally collect everyone's feedback into a single place so it can become actionable.

With this new functionality, views and data are extracted from your drawing, stored in the cloud, and a shareable link generated. You can then send that link to stakeholders who will be able to view, review, measure, comment, and markup the view you've shared with them.

Intended as a tool to streamline collaboration during the design process, shared views automatically expire after 30 days, but you have the option to extend or terminate links whenever necessary.

Updated 4K-compliant icons. A refresh of icon imagery for ribbon and menu elements features autosensing for 4K monitor users.

So the new features in AutoCAD 2019 are actually quite limited, and more collaboration- and web-centric than CAD-centric. While the 2D graphics speed increase will be welcomed by users with large drawings, and the 4K icons welcomed by those with the hardware to support them, the rest of the changes will only be appreciated by those who are collaborating via web/mobile methods.

Bottom line: If you use AutoCAD as a desktop application and don't utilize any collaborative or web features, you won't notice much difference in the new version.

1.3 Creating an AutoDesk account

AutoCAD is a computer-aided design software developed by AutoDesk Inc. that is a very thorough and professional software design suite with the ability to generate sophisticated results. You must create an account on their website to use Autodesk software.

This software is quite expensive, because it is intended for 3D design professionals. If you want to enter CAD in general, there are also some free alternatives, which are listed here.

Name	Level	OS	Price	Formats
Photoshop CC	Beginner	Windows and Mac	142 €/year	3ds, dae, kmz, obj, psd, stl, u3d
TinkerCAD	Beginner	Browser	Free	123dx, 3ds, c4d, mb, obj, svg, stl

LibreCAD	Beginner	Windows, macOS and Linux	Free	dxf, dwg
3D Slash	Beginner	Windows, Mac, Linux, Raspberry Pi or Browser	Free, 24€/year Premium	3dslash, obj, stl
SculptGL	Beginner	Browser	Free	obj, ply, sgl, stl
SelfCAD	Beginner	Browser	Free 30-day trial, $9.99/month	stl, mtl, ply, dae, svg
SketchUp	Intermediate	Windows and Mac	Free, 657€ Pro	dwg, dxf, 3ds, dae, dem, def, ifc, kmz, stl
FreeCAD	Intermediate	Windows, Mac and Linux	Free	step, iges, obj, stl, dxf, svg, dae, ifc, off, nastran, Fcstd
OpenSCAD	Intermediate	Windows, Mac and Linux	Free	dxf, off, stl
MakeHuman	Intermediate	Windows, Mac, Linux	Free	dae, fbx, obj, STL

Meshmixer	Intermediate	Windows, Mac and Linux	Free	amf, mix, obj, off, stl
nanoCAD	Intermediate	Windows	Freemium, $180/year	sat, step, igs, iges, sldprt, STL, 3dm, dae, dfx, dwg, dwt, pdf, x_t, x_b, xxm_txt, ssm_bin
DesignSpark	Intermediate	Windows	Freemium, $835 (All Addons)	rsdoc, dxf, ecad, idf, idb, emn, obj, skp, STL, iges, step
Clara.io	Intermediate	Browser	Free, Premium features from $100/year	3dm, 3ds, cd, dae, dgn, dwg, emf, fbx, gf, gdf, gts, igs, kmz, lwo, rws, obj, off, ply, pm, sat, scn, skp, slc, sldprt, stp, stl, x3dv, xaml, vda, vrml, x_t, x, xgl, zpr
Moment of Inspiration (MoI)	Intermediate	Windows and Mac	266 €	3ds, 3dm, dxf, fbx, igs, lwo, obj, skp, stl, stp and sat
AutoCAD	Professional	Windows and Mac	1400 €/year	dwg, dxf, pdf

Blender	Professional	Windows, Mac and Linux	Free	3ds, dae, fbx, dxf, obj, x, lwo, svg, ply, stl, vrml, vrml97, x3d
Cinema 4D	Professional	Windows, macOS	$3,695	3ds, dae, dem, dxf, dwg, x, fbx, iges, lwf, rib, skp, stl, wrl, obj
3ds Max	Professional	Windows	3.241,70 €/ year, Educational licenses available	stl, 3ds, ai, abc, ase, asm, catproduct, catpart, dem, dwg, dxf, dwf, flt, iges, ipt, jt, nx, obj, prj, prt, rvt, sat, skp, sldprt, sldasm, stp, vrml, w3d xml
ZBrush	Professional	Windows and Mac	400€ Educational License, 720€ Single User License	dxf, goz, ma, obj, stl, vrml, x3d
modo	Professionals	Windows, macOS, Linux	$1799	lwo, abc, obj, pdb, 3dm, dae, fbx, dxf, x3d, geo, stl

Onshape	Professional	Windows, Mac, Linux, iOS, Android	2.400 €/year, free and price reduced business version available	sat, step, igs, iges, sldprt, stl, 3dm, dae, dfx, dwg, dwt, pdf, x_t, x_b, xxm_txt, ssm_bin
Poser	Professionals	Windows, Mac	Standard $129.99, Pro $349.99	cr2, obj, pz2
Rhino3D	Professional	Windows and Mac	495€ Educational, 1695€ Commercial	3dm, 3ds, cd, dae, dgn, dwg, emf, fbx, gf, gdf, gts, igs, kmz, lwo, rws, obj, off, ply, pm, sat, scn, skp, slc, sldprt, stp, stl, x3dv, xaml, vda, vrml, x_t, x, xgl, zpr
Mudbox	Professional	Windows and Mac	85 €/year	fbx, mud, obj
Solidworks	Industrial	Windows	9.950 €, Educational licenses available	3dxml, 3dm, 3ds, 3mf, amf, dwg, dxf, idf, ifc, obj, pdf, sldprt, stp, stl, vrml
Inventor	Industrial	Windows and Mac	2,060 €/year	3dm, igs, ipt, nx, obj, prt, rvt, sldprt, stl, stp, x_b, xgl

Fusion 360	Industrial	Windows and Mac	499.80 €/year, Educational licenses available	catpart, dwg, dxf, f3d, igs, obj, pdf, sat, sldprt, stp
CATIA	Industrial	Windows	7.180 €; Educational licenses available	3dxml, catpart, igs, pdf, stp, stl, vrml

But there is good news: you can get AutoCAD and all AutoDesk products for three years if you are a student. To activate your student license, enter your educational email address for registration. If you are not lucky enough to receive a student discount, you can still activate a 3-month free trial for all Autodesk products.

1.4 Install the software

Once you have completed the registration process, you should download the AutoCAD installer. Run the downloaded file. All this will download and open the installation wizard. If necessary, you can change the installation directory, choose the components to install or install or install AutoCAD immediately. Then the AutoCAD download is started.

Installation process...

1. Double click on installation file, and then click 'Yes' to complete the installation. If prompted, read and accept the license agreement for your country or region.

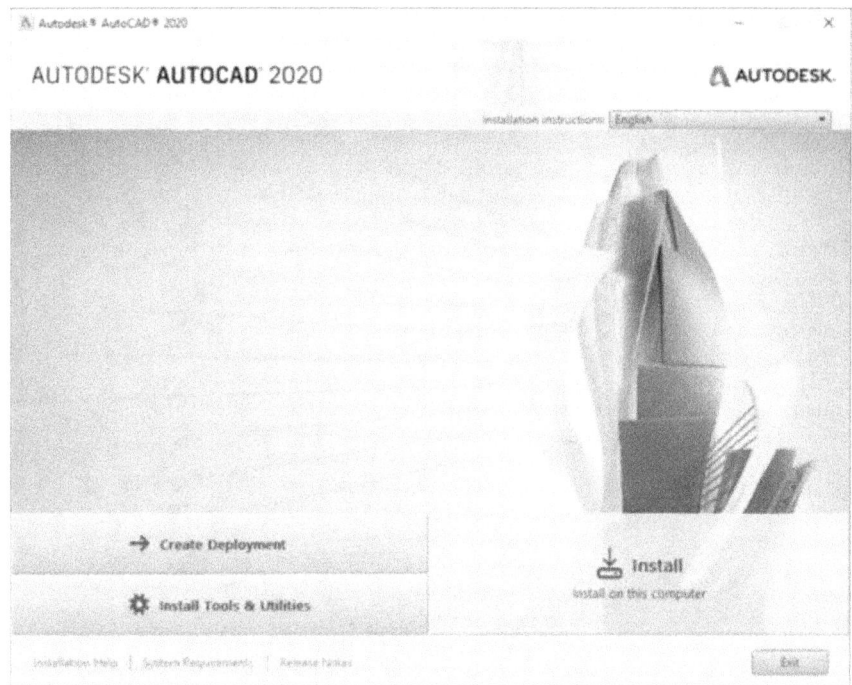

2. Do the following and click Install:

- Select the products or components to install.
- Specify where the installed files will be located.

If your product displays a Product Language drop-down menu, you can choose the language appropriate for the person who will be using it. This process can take several minutes. Click on 'Install'

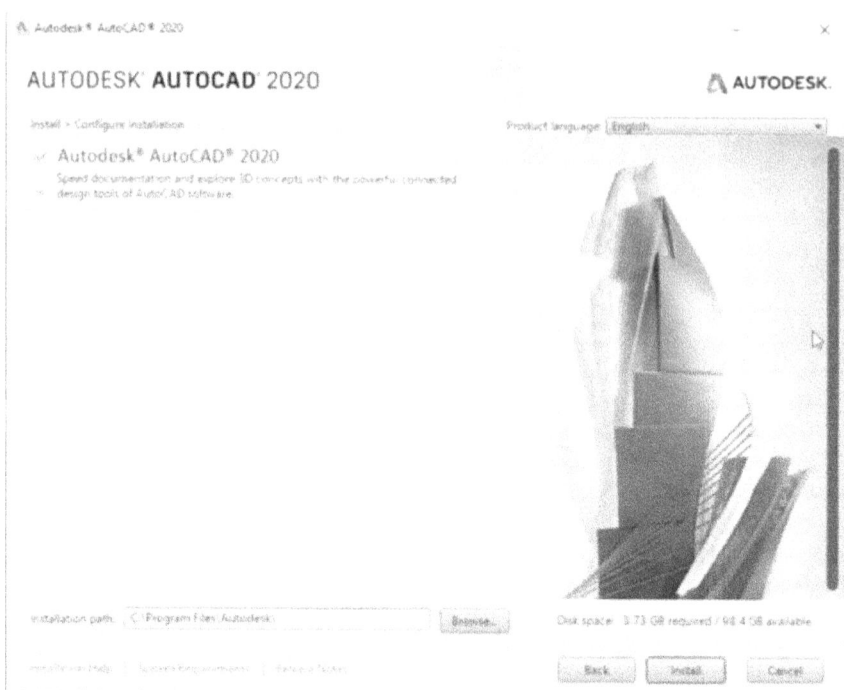

3. When the installation is done you'll see a list of the installed software components. Click Finish to close the installer.

AUTODESK **AUTOCAD** 2020

⚠ AUTODESK.

Install > Installation Complete

You have successfully installed the selected products.
Please review any product information alerts.

⚙ ✓ Autodesk® AutoCAD® 2020

Speed documentation and explore 3D concepts with the powerful connected
design tools of AutoCAD software.

✓ Autodesk Desktop App

Installation Help | System Requirements | Release Notes

[Finish]

CHAPTER 2 GET AROUND IN AUTOCAD

Welcome to the AutoCAD's World. This AutoCAD tutorial will teach you the basics of using AutoCAD and create your first objects. AutoCAD is a robust tool for creating 2D and 3D objects, like architectural plans and constructions or engineering projects. It can also generate files for 3D printing. If you want to start this AutoCAD tutorial for beginners, you should have about an hour to do so.

In this first chapter, I'll introduce you to commands and AutoCAD's user interfaces. But first, you have to know why CAD software is now replaces traditional pencil drawing and now you don't have to use that big, drafting table to draw an advanced drawing.

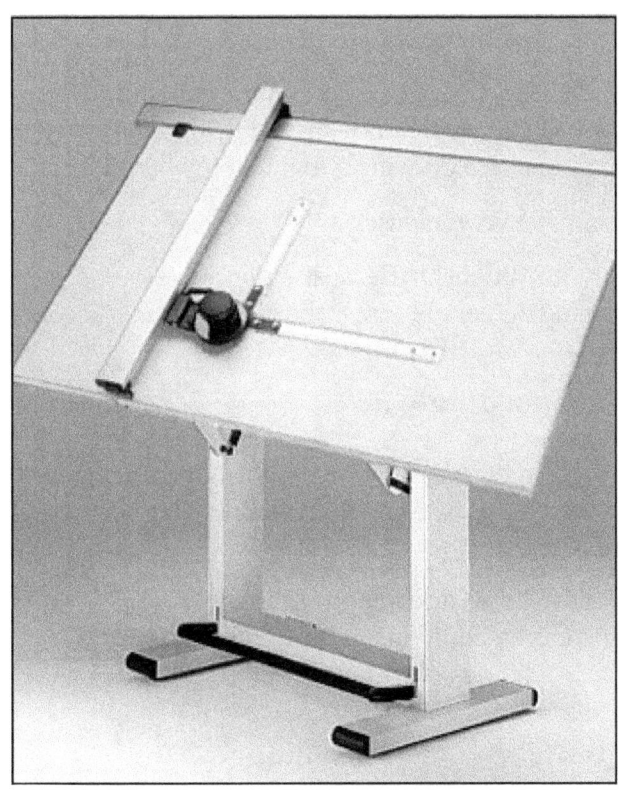

Pic 2.1 Most of current engineering drafter do not have this "ancient" drafting table

2.1 Why AutoCAD?

These are features of CAD's software that make drawing with software is better:

- Precision: You can draw a line, the arc, and other forms with incredible precission. Accuracy in AutoCAD is 14 decimal point.

- Modifiable: An advanced drawing created a long time ago can be modified again to draw a new drawing. While old drawing in pencil/pen cannot be updated and you have to create the new drawing from scratch.

- Clean: You don't have to own eraser to draw a drawing.

- Efficiency: You can create more drawing in the same time, and you can create drawing faster. Especially when you need repetition, like drawing a multi-storey building, or floor tiles.

- Popular: Everyone uses it.

- Easy to Publish: Because it's digital, you can give the drawing on people across the globe just in a second.

2.2 XY Coordinate

All objects in AutoCAD are exactly positioned. For this reason, you need to understand how AutoCAD defines the position with simple X, Y coordinates.

AutoCAD has the World Coordinate System (WCS). For the 3D drawing, there is an additional Z axis.

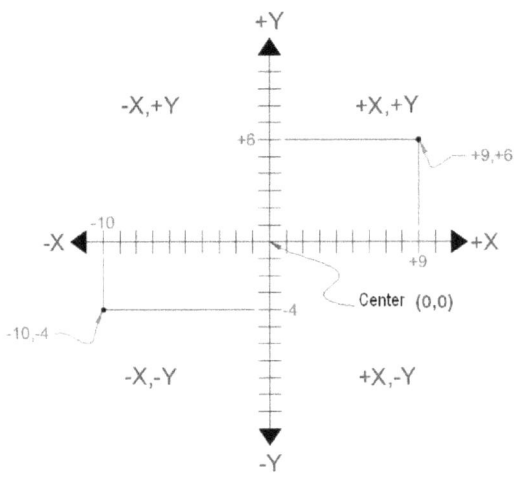

Pic 2.3a Simple XY WCS coordinate used in AutoCAD

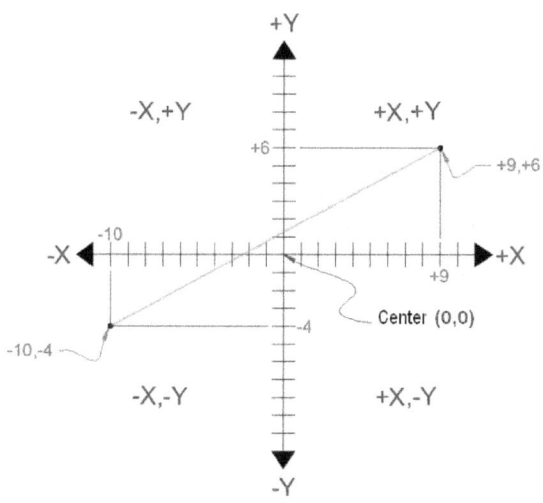

Pic 2.3b A line from -10,4 to e +9, +6

The AutoCAD has x,y point indicating where the point object is located. It also has an origin or center point (0.0) where all object positions use this point as their first reference.

See the image above to see the AutoCAD x, y coordinates and how to draw a line between two coordinates. For example, the xy coordinate 9.6 represents x = 9 and y = 6.

Coordinate (-10,-4) means x = 10 units negative (left side) and y = 4 units negative (bottom).

In some cases you don't know the exact starting position. You just know that you want to draw to the next point relative to that position. You can use the relative coordinate by adding the @ (SHIFT + 2) icon to tell AutoCAD that the next point is relative to the last point..

Here are some important points about X, Y coordinate.

- The absolute point is the exact position of a point, relative to 0,0.

- The relative point is relative to the last point.

2.3   Angle in AutoCAD

AutoCAD also has angle to draw. Here's how to specify the angle in AutoCAD:

✓ The X positive is the 0 degree.

✓ Counter clock wise is positive

✓ Clockwise is negative.

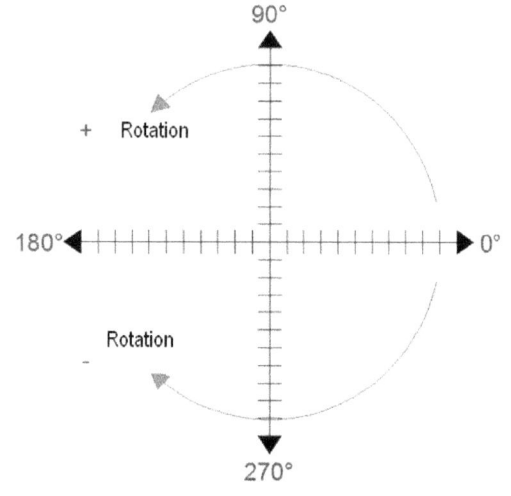

Pic 2.3 Angle measurement in AutoCAD

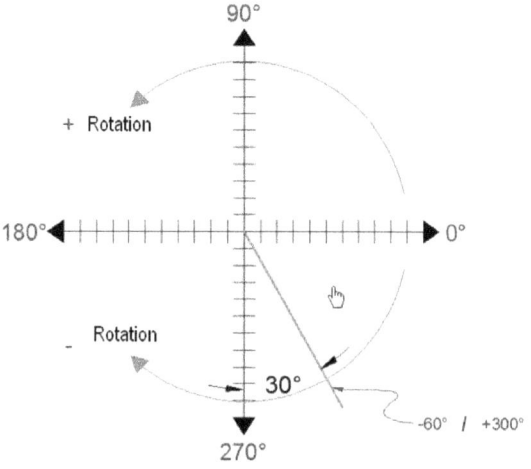

For example 90 degrees = Y positive.

You can measure angle based on other angle.

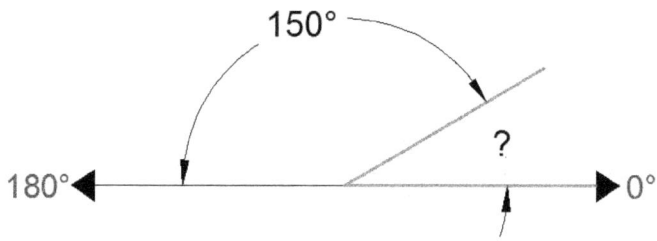

Pic 2.5 Calculating an angle from another angle

From above examples, some important notes:

- 0 degrees on hour = 3 position.

- 180 degrees on hour = 9 position

2.4   Inserting Point in AutoCAD

Here are three methods to insert point in AutoCAD:

- Absolute coordinate: Just insert the xy point relative to the center point (0,0). Insert x value first, and then y value.

- Relative coordinate: Insert by adding prefix @ so you enter @X,Y. This will put point x,y points relative to the last position.

- Polar coordinate: Insert by using template @D<A. Which D is the length and A is the angle: For example @10<90 will draw a line with length = 10 units and 90 degrees direction.

Notes:

- The three methods are the only methods for inserting point in AutoCAD, there are no other methods to draw AutoCAD. X value has to be inserted first, then Y value.

- Don't forget the '@' symbol when you insert relative value. All mistakes in inserting input will generate unintended results.

- If you want to do checking, click F2, and then click F2 again

2.5   AutoCAD's User Interfaces

In the second step of this AutoCAD Tutorial, you will learn how to interact with the workspace. When you run AutoCAD program for the first time, you can see AutoCAD's window like the picture below:

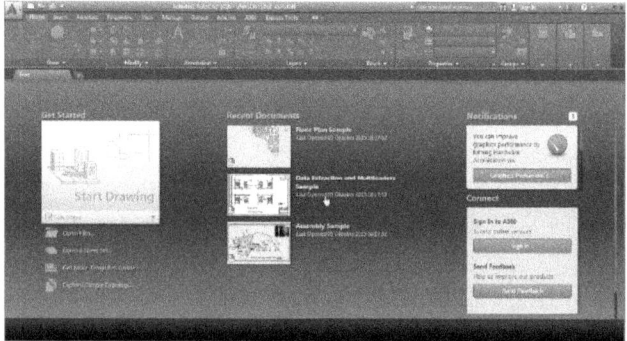

Pic 2.6 Start window

To draw a new file, click Start Drawing, an AutoCAD's user interface for the drawing will be displayed.

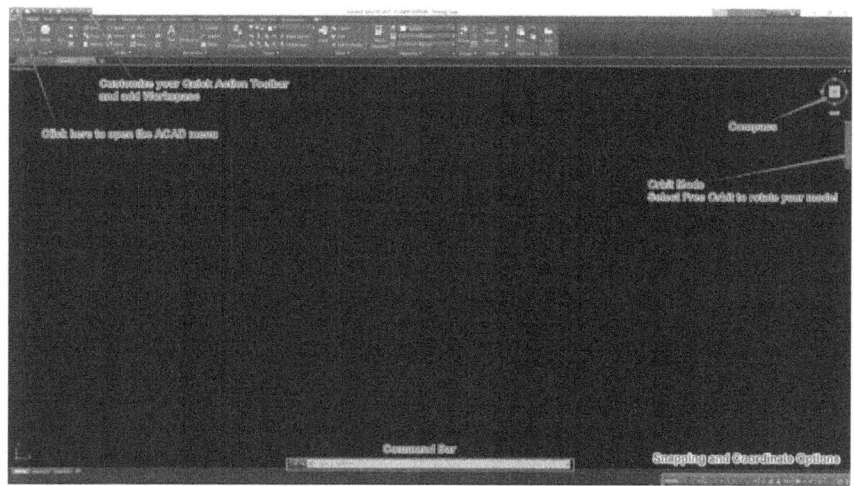

Pic 2.7 Drawing interface

When opening the software for this AutoCAD tutorial, click on "Start Drawing" to open a new file or project. By doing this, you opened the "DrawSpace."

At first, you need to customize your Quick Action Toolbar and add "Workspace" by clicking on it. Change now the new Toolbar "Drafting and Annotation" to "3D Modelling". This will allow the use of all the Sketch and 3D Tools you need to design your first sketch and 3D Object.

2.5.1 Change Units in AutoCAD

If you want to change units to the metric system you're used to, click on the big red A in the top left corner. This will open the AutoCAD menu. Select "Drawing Utilities" > "Units". Change the Insertion Scale to Millimeters.

2.5.2 Explanation of the Workspace

- The Command Bar

The command bar is located at the bottom of the DrawSpace (see figure above). You can enter the commands either simply by typing them into the command bar. It shows you contextually the options you have received for the given command. Highlighted letters are abbreviations for these options.

By entering the corresponding letter and pressing "Enter", the desired option is activated directly. It also lists the order of the steps you need to perform to execute the command correctly and display tips.

- Orientation in AutoCAD

In the upper right corner of your DrawSpace you will see a compass. It is set to "top view" by default. Move the mouse pointer over it and you will see a small house symbol. Click on it to get to the isometric view. Now you will see a 3D-Cartesian coordinate system with three axes in the middle of your DrawSpace. The x-axis is red, the y-axis green and the z-axis blue.

The compass has also been extended by a cube. You can click on the faces, edges and corners of the cube to open the desired view. To move the DrawSpace, click the Hand icon or move with the mouse wheel held down. If you want to orbit your DrawSpace, click Orbit on the right toolbar. Click and hold the DrawSpace to rotate around the center of the coordinate system by moving the mouse. You can also do this by holding down the Shift key and the mouse wheel. If you want to orbit a specific point, select "Free Orbit" by clicking on the expansion arrow.

To move the DrawSpace, click on the "Hand" icon or move with the mouse wheel held down. With the Zoom Extends option, you can fit all your created objects and sketches into your field of view.

At the moment you have nothing to revolve around, so hope for the next step of this AutoCAD tutorial to start sketching!

2.5.2 Ribbon

When you on drawing page, buttons in the ribbon will be enabled. The ribbon interface is similar with MS Office interface so you will be familiar and makes drawing process easier.

In HOME tab, you'll see buttons to Draw, and Modify drawing.

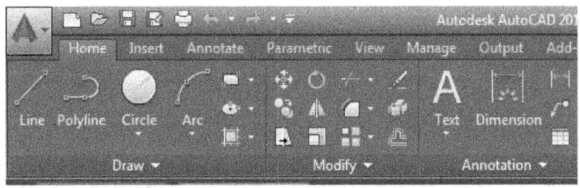

Pic 2.8 Draw and Modify

Still in Home tab, there are Annotation and Layers, the Annotation box used to give annotation to your drawing, eg: text, dimension, etc. The Layers box is to insert layers to your drawing. The user can add a layer to overlay the drawing.

Pic 2.9 Annotation and Layers boxes

There are **Block**, **Properties**, and **Groups** boxes. **Block** contains buttons to block more than one object to become a single object. The **Properties** box used to manage the properties of an object. **Groups** to group or ungroup objects.

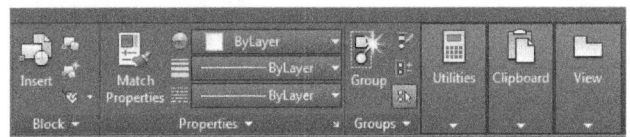

Pic 2.20 Block, Properties and Groups

Insert box is used to insert many kind of objects, from Block, Definition, Reference, Point Cloud and Import.

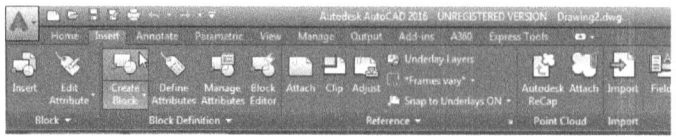

Pic 2.21 Insert tab

Annotate tab used to insert more detail anotate, from texts, dimensions, leaders, etc.

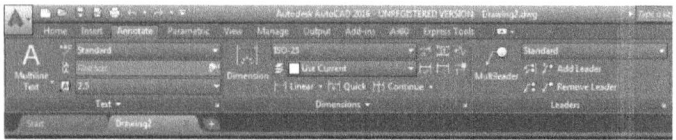

Pic 2.22 Tab Annotate

The parametric tab has buttons that used for managing geometric and dimensional drawing.

Pic 2.23 Parametric tab

View tab is used to modify the user interface of AutoCAD. You can manage the Viewport, Palettes, and Interface.

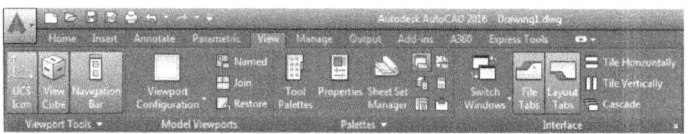

Pic 2.24 View tab from ribbon

Manage tab, used to create macro used to record your action. You can do coding in macro.

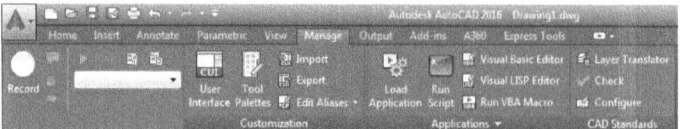

Pic 2.25 Manage tab

Output tab used for exporting and printing your drawing to paper or other forms.

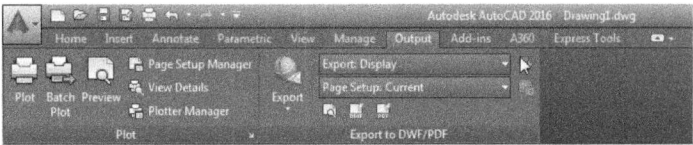

Pic 2.26 Tab Output

In Add Ons tab, you can manage add-on applications.

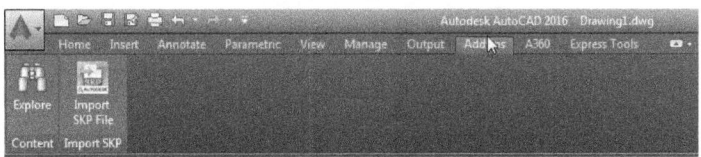

Pic 2.27 Add On application

A360 tab consists of buttons that enable you to use online features of AutoCAD.

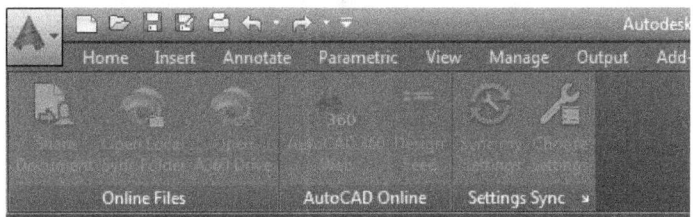

Pic 2.28 Cloud saving function

Express Tools tab can be used to manage blocks, texts, modifying objects and layout.

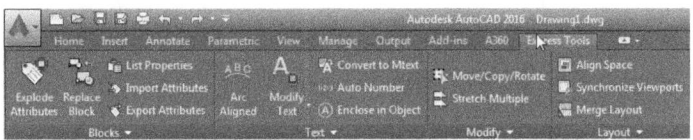

Pic 2.29 Tab Express Tools

Ribbon in AutoCAD can be minimized, just click 2x on the ribbon tab.

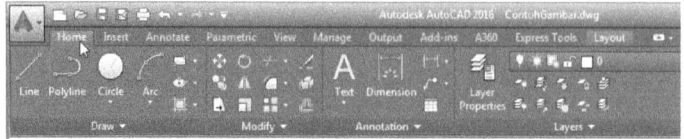

Pic 2.30 Double click on ribbon tab

The ribbon will be minimized.

Pic 2.31 Buttons in ribbon minimized

If you click twice again, the buttons are hidden, and the ribbon only displays the texts.

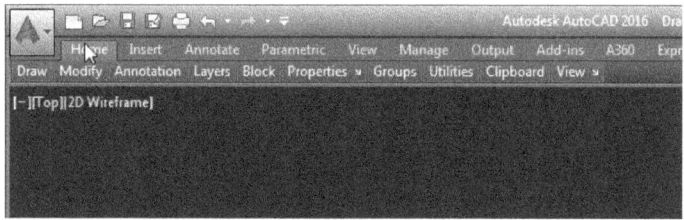

Pic 2.32 Ribbon's buttons hidden

2.5.2 Menus

Main Menus can be opened by clicking A button on the top left of your AutoCAD window:

Pic 2.33 Main menu of AutoCAD

On the menu box above, you can see Sethe arch Command textbox that makes finding commands easier. Just enter the name of the command, and AutoCAD will autocomplete it for you.

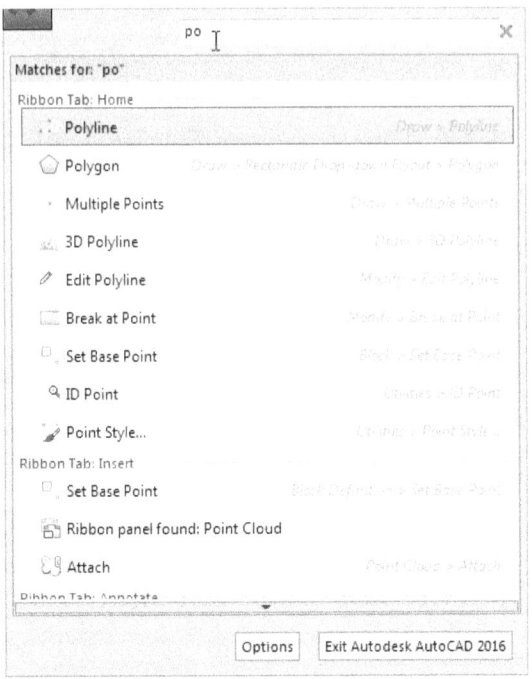

Pic 2.34 Inserting command name in AutoCAD

This menu can be accessed from all parts of the workspace. Menus in the main menu are:

1. New: To draw a new drawing, from template, or create sheet set that manages drawing layouts, paths, and project data.

Pic 2.35 New menu

2. Open : To open existing drawing.

Pic 2.36 Menu Open

3. Save: Save existing drawing changes, if the drawing hasn't saved before, it will save to a new file.

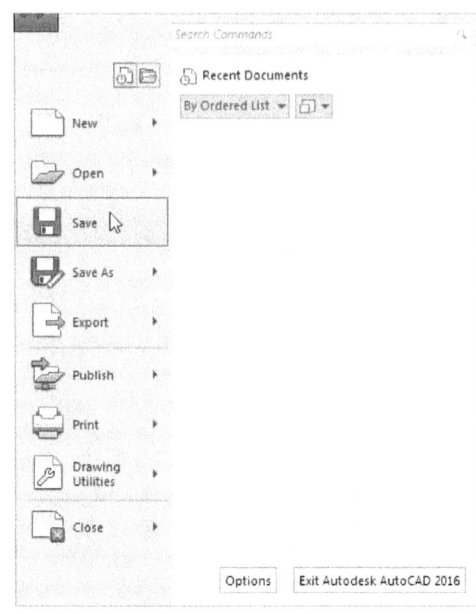

Pic 2.37 Save menu

4. Save As: Save existing drawing to a new file.

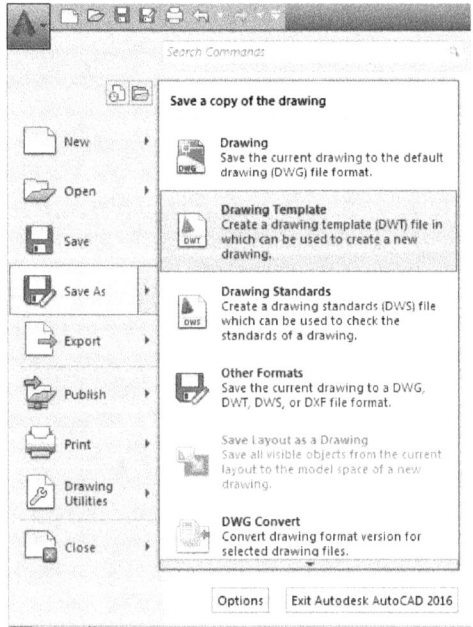

Pic 2.38 Menu Save As

5. Export: Save drawing to other file formats, such as Design Web Format (DWF), PDF, and other CAD files.

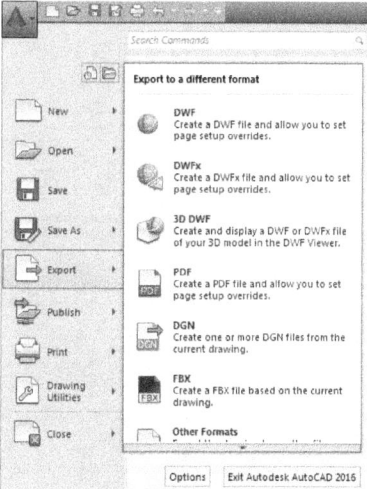

Pic 2.39 Export Menu

6. Publish: Send 3D model to 3D printing service, or create the archived sheet set (AutoCAD LT doesn't support 3D.) etc.

Pic 2.30 Publish menu

7. Print: Printing single drawing, or batch-plot. You can also setup the page and style plot.

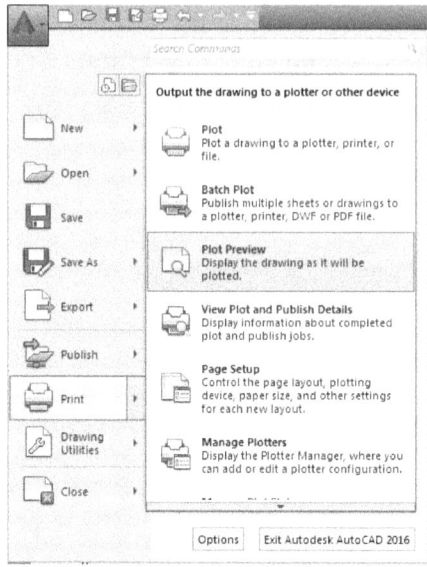

Pic 2.31 Print menu

8. Drawing Utilities: Setting file properties, or drawing unit, doing purging on unused blocks, doing auditing or recovering damaged drawing.

9. Close: Closing existing drawing, if the drawing already modified and hasn't been saved, this will generate Save confirmation box.

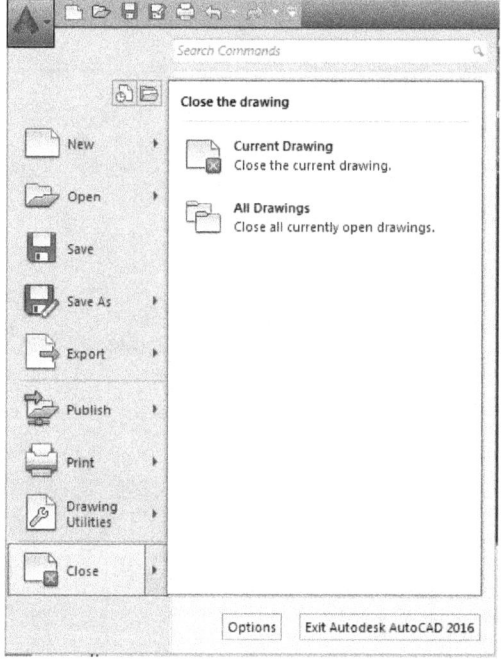

Pic 2.33 Close menu

2.6   Open Drawing

You can open drawing file to display on your AutoCAD using steps below:

2. Click on AutoCAD icon to display AutoCAD:

2. Click Open > Drawing menu.

Pic 2.34 Click Open > Drawing

3. **Select File** window appears, choose the file you want to open, click **Open**.

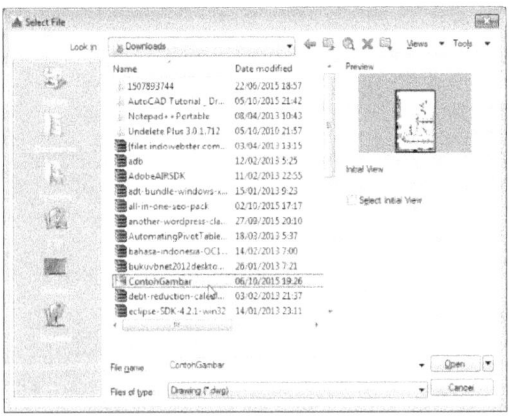

Pic 2.35 Choosing the file to open

4. The drawing will be opened.

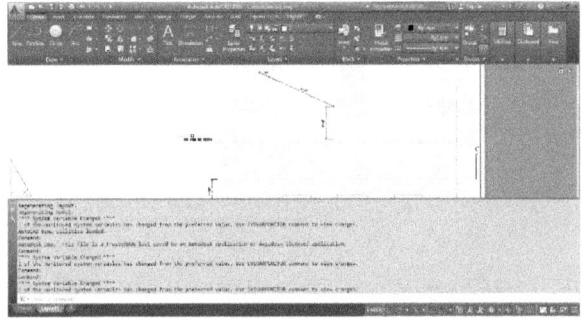

2.36 Drawing opened in AutoCAD

5. AutoCAD can display more than one drawing. Each drawing will be opened as MDI (multiple document interface) windows.

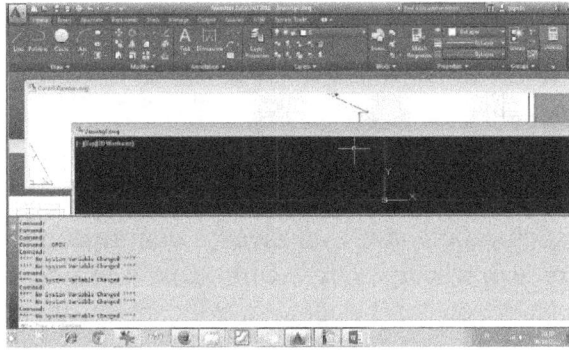

Pic 2.37 AutoCAD may open more than one project

2.7   Close Drawing

A drawing that doesn't need to be edited further, close it using steps below:

1. Click AutoCAD icon to open the main menu.

2. Click **Close** > **Current Drawing** menu to close the active drawing.

3. Or click **Close** > **All Drawing** to close all drawing.

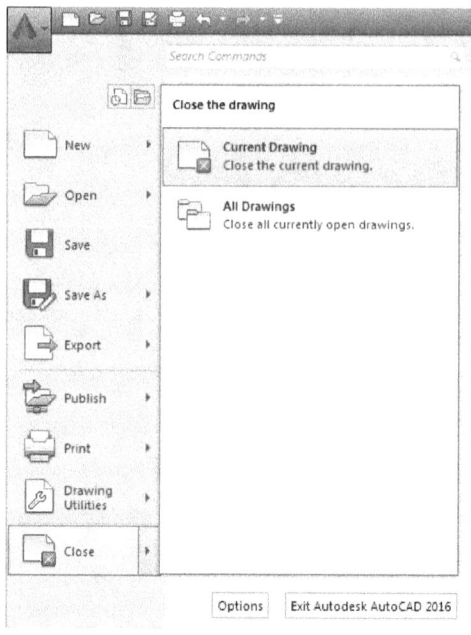

Pic 2.38 Close menu to close the drawing

4. If your modification hasn't yet saved, a confirmation window will appear and ask whether you want to save it or not. Click Yes to save and No if you don't want to save the modification.

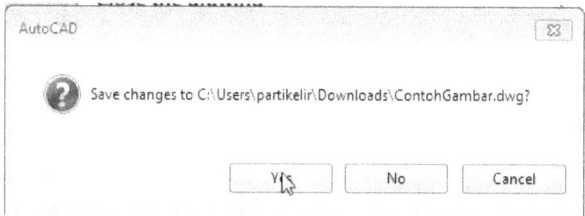

Pic 2.39 Confirmation window

2.8   Export as PDF

PDF (portable document format) is the most pervasive format used in the design world. AutoCAD can export it's drawing directly to pdf without the third-party software or add-in.

Look at steps below to export your drawing as PDF:

1. Click on AutoCAD icon.

2. Click on **Export > PDF** menu.

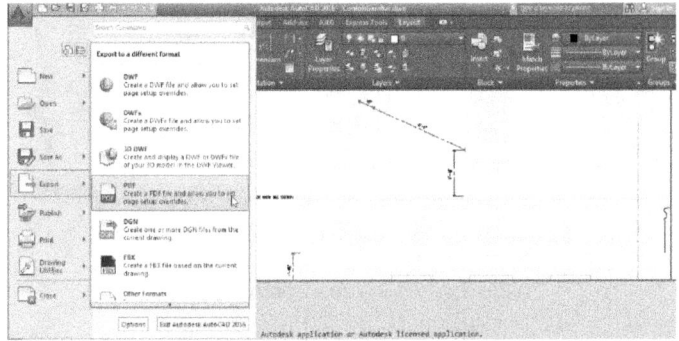

Pic 2.40 Menu Export > PDF

3. **Save As PDF** window emerges, choose a filename for the new pdf file in **File name** textbox. And click **Save**.

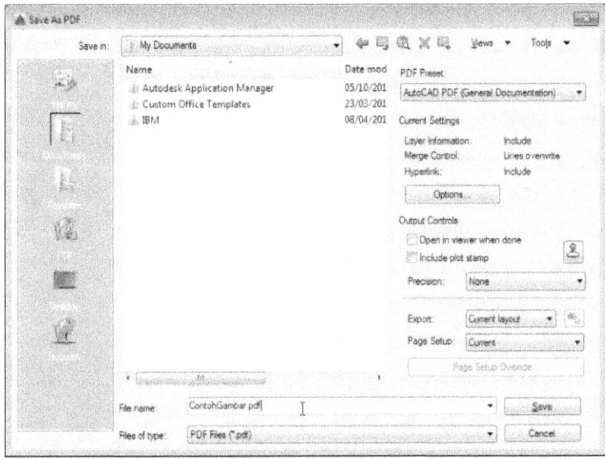

Pic 2.41 Inserting file name for pdf file

4. The file will be inserted as PDF.

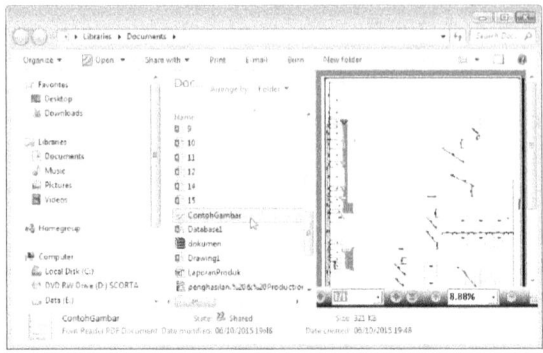

Pic 2.42 PDF file already created

CHAPTER 3 DRAWING IN 2D

This chapter will explain important commands that you can use to draw 2 dimensional drawings in AutoCAD. Drawing in 2D is the basis of the AutoCAD drawing.

3.1 Set up Snapping

When sketching with AutoCAD, you can make use of its Snapping option. To enable Grid Snap, simply press F9 on your keyboard or click on enable "Snap to Drawing Grid" in the bottom right-hand corner. By opening Snap Settings, you can adjust the Drawing Grid as well as the accuracy of Grid Snap.

By pressing F3 or clicking on Object Snap, you can activate clipping to corners, lines, points, midpoints and many more. Edit the object snapping to your current drawing goals. If you have problems with entering coordinates or sketching, try turning snapping on or off and try not to use Grid and Object snap simultaneously. This tool is useful to draw sketches fast and to prevent holes in your sketch.

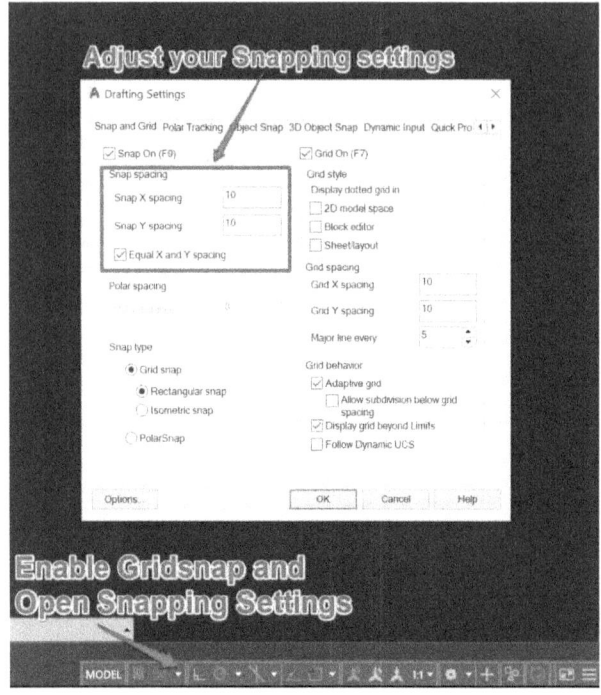

Pic 3.1 Set up Snapping

3.2 Create 2D Drawing

There are lots of 2d drawing types you should understand. From Line to Donut. I'll show you how to draw 2d using those types of drawing.

3.2.1 Drawing a line

Line is the basic type of drawing. It's a straight line that connects two dots. You can draw a line by following steps below:

1. Click **Line** button in **Home** > **Draw** ribbon.

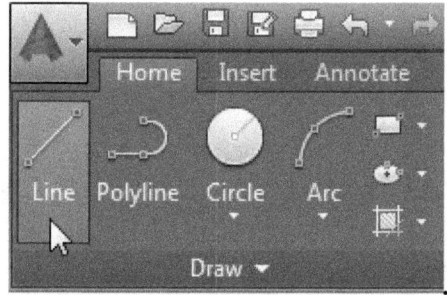

Pic 3.2 Click Line button in Home > Draw

2. Or you may type "line" in the command prompt.

3. Command prompt appears:

```
LINE from point:
```

4. You can insert with absolute coordinate or click on the drawing.

```
Prompt: To point:
```

5. Insert the second point location.

6. Before creating LINE, limits your workspace by inserting LIMITS command.

```
Command: LIMITS
Reset Model space limits:
```

7. Set the bottom-left limit to 0,0.

```
Specify lower left corner or [ON/OFF] <0.0000,0.0000>: 0,0
```

8. Then specify the top right limit to 100,100. This will make creating picture easier, because the canvas for this tutorial is from 0,0 to 100,100.

```
Specify upper right corner <420.0000,297.0000>: 100,100
```

9. Then type the line to start creating line, specify the first point to 10,10.

```
Command: LINE
Specify first point: 10,10
```

10. When you move the pointer, you'll see that the first point of line glued to 10,10, and you still can move the mouse pointer.

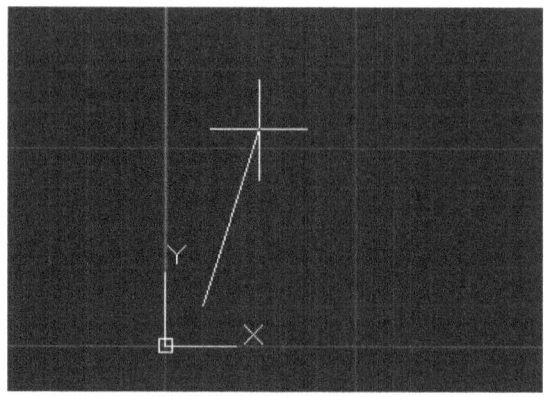

Pic 3.3a The first point of line glued to 10,10

11.   You can change the pointer right or left.

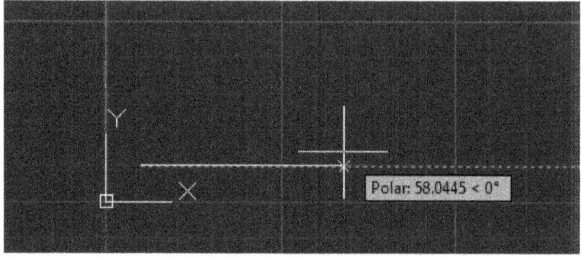

Pic 3.3b Pointer mouse still can be moved

12.   For the next point, choose 50,50. You can see the line glued to 50,50.

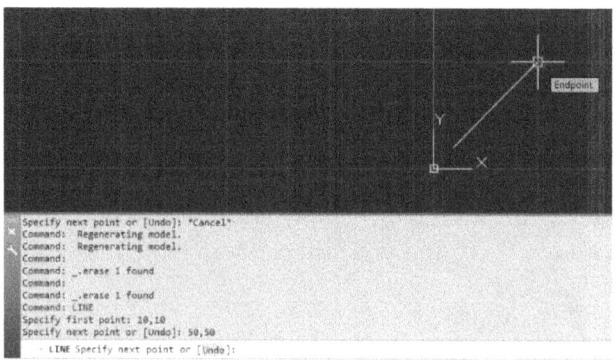

Pic 3.4 Line connected from 10,10 to 50,50

13. Click Enter, a line will be created, and the pointer released from the line.

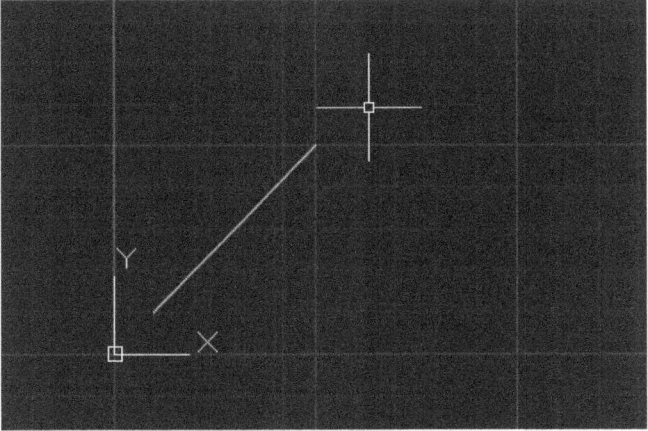

Pic 2.5 Line created, and the pointer mouse released

14. All command line texts in this tutorial:

```
Command: LINE
Specify first point: 10,10
Specify next point or [Undo]: 50,50
Specify next point or [Undo]:
```

In the next tutorial, we'll draw a line using relative coordinate, you can see the steps below:

1. Type Line

```
Line
```

2. Specify the first point = 10,10.

```
Specify first point: 10,10
```

3. Specify next point @50,25 relative from the first point.

```
Specify next point or [Undo]: @50,25
```

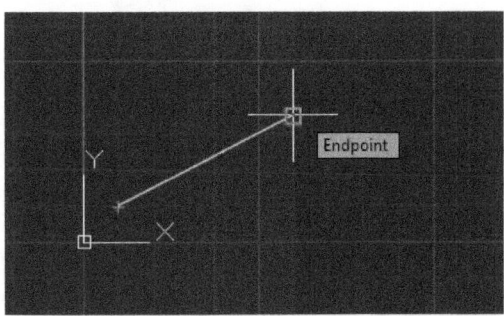

Pic 3.6 Specify second point using relative coordinate

4. Click Enter, line will be created.

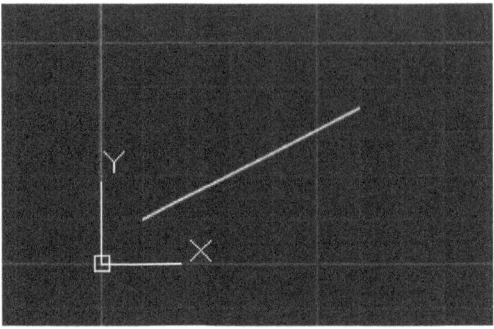

Pic 3.7 Line created using relative coordinate for the second point

5. To delete line, click on the line to select the line first. Selected line will become dotted line.

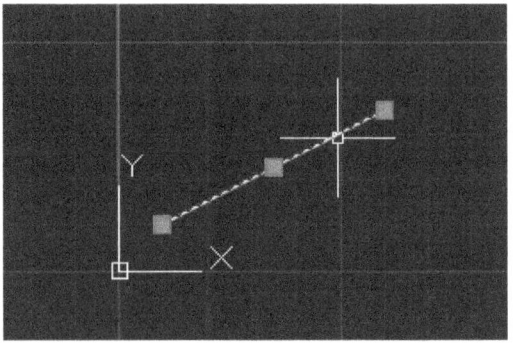

Pic 3.8 Selected Line become dotted

6. Click **Delete** button on your keyboard, or right-click and click **Erase** menu.

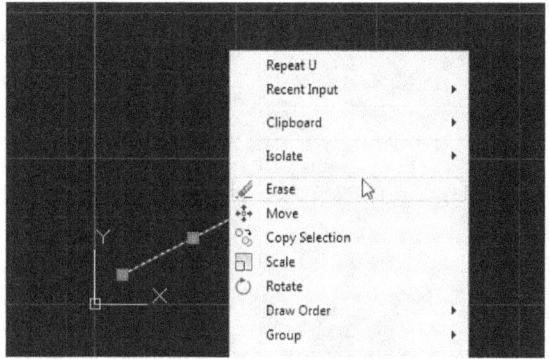

Pic 3.9 Erase menu to delete selected line

The third tutorial is using angle coordinate. Here's the method to draw a line using angle coordinate:

1. Insert line command, and specify the first point = 10,10.

```
Command: LINE
Specify first point: 10,10
```

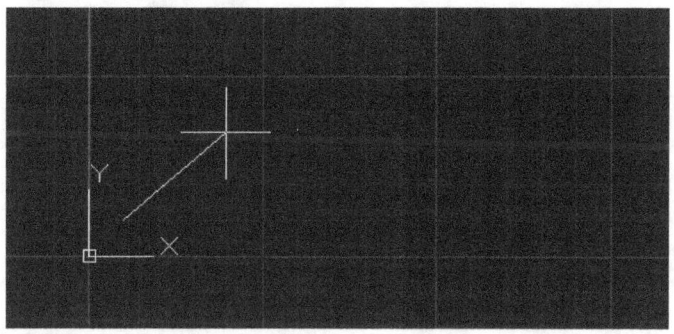

Pic 3.20 Specify the first point = 10,10

2. Then Specify second point 50 units from the first point, and with <45 degrees. Click Enter:

```
Specify next point or [Undo]: @50<45
Specify next point or [Undo]:
```

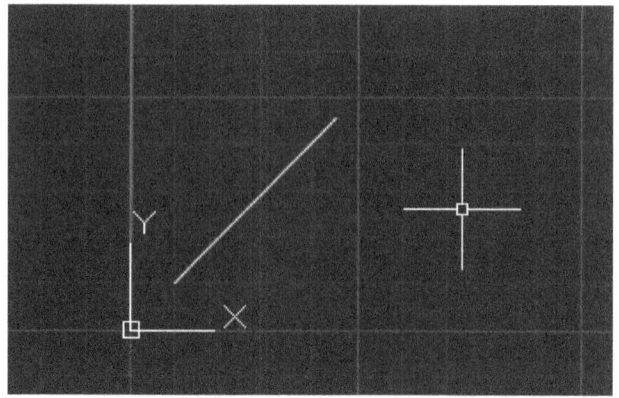

Pic 3.21 Drawing line by degree coordinate

✓ **Exercise Drawing a Line**

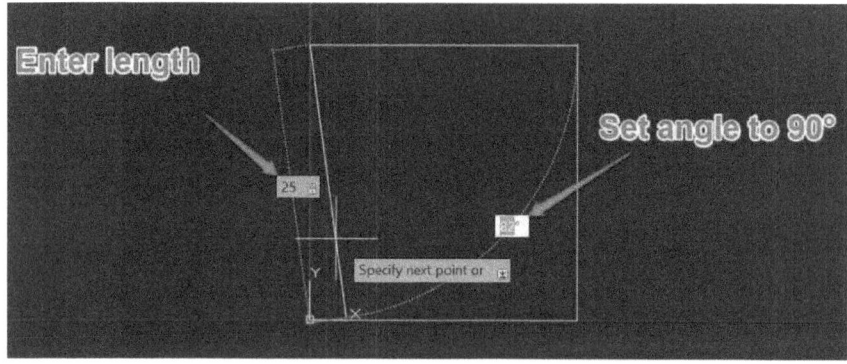

To create your first sketch, select Top View with the compass. Disable Grid Snap by pressing F9. Now type "line" and press Enter. This will enable the Line command.

With AutoCAD, you can simply type in the first letters of any command. The software will autocomplete or show any available commands. When you have entered the line command, it asks you to specify the first point. You can now either select a random point in your DrawSpace or enter the coordinates. Enter 0 for X-Coordinate, change to Y-Coordinate by pressing Tab, enter 0 as well and confirm your coordinates by pressing Enter. You have now selected the center of the coordinate system as Start.

Now move your mouse to the positive side of the X-Axis. You can now see how the coordinate input changed to Polar coordinates. Enter 25 for the length of the line by pressing Tab you can switch to the angular input. Try sketching a square for starting. When you have returned to the center, press Escape to end the line command.

✓ ***Exercise Drawing a Line***

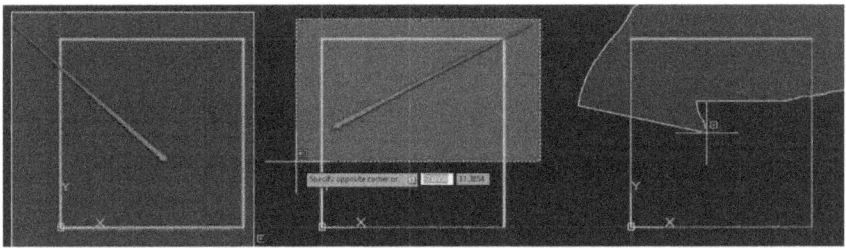

To select objects, you can click on them. Unselect by holding down the "Shift" key and clicking again. Select multiple objects by left-clicking and moving from left to right. This will select all objects fully enclosed within the blue rectangle. When you drag from right to left, you will select all objects touched by the green rectangle. Click again to confirm the selection. Clicking and holding the left mouse button will enable the lasso, which lets you select a random shape.

3.2.2 Drawing Polyline

Polyline is multiline, more than one lines that composed by line and the arc segments. See picture below for polyline example:

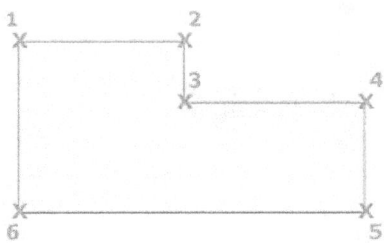

Pic 3.22 Polyline example

Some notes on polyline:

- Specify start point: similar to LINE command, specify the first point or initial point.

- Next:

```
polyline, line, or the arc.
Specify next point
```

- If you choose the second point, you'll create straight line.

- If you enter other option, for example the arc, you'll make an the arc.

There are some prompts related to line and the arc:

- Close : connects first segment and the last segment to draw a closed polyline.

- Halfwidth: half width of the segment, from the center to outer.

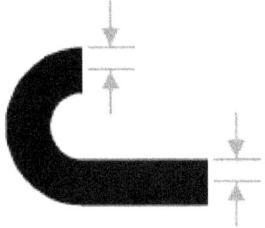

Pic 3.23 Halfwidth

- Width: The width of next segment.

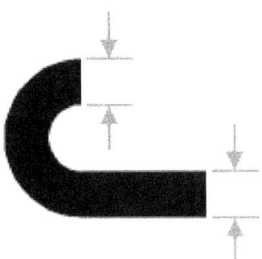

Pic 3.24 Width

- The first part of width will equal the last width. The last width will be uniform to all segments until you change another width. The first part and the end part of width is similar to the width at the middle of the segment.

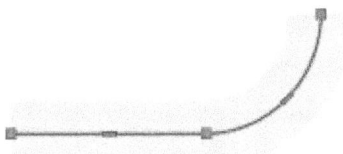

Pic 3.25 Width in line and the arc

- At the intersection of a polyline, there will be a bevel.

Pic 3.26 Beveling in polyline

- Undo will erase the last segment added.

Some arguments in Line-Only prompt:

- The arc: creating the arctangent from the previous segment.

- Length: Creating segment with length = next segment. If the next segment is an the arc, the new segment will be tangent from the arc segment.

Pic 3.27 Length

Some arguments in the arc-only prompts:

- Endpoint of the arc: Completing the arc segment. Tangent from the previous polyline segment.

- Angle: Specifying the angle from the arc segment from center point. If positive = counter clockwise, if negative = clockwise.

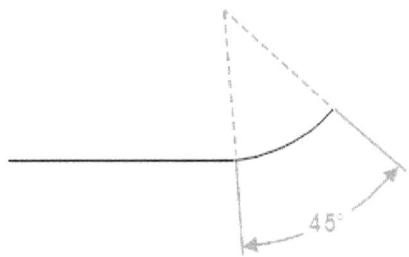

Pic 3.28 Angle

- Center: Specifying the arc segment based on center point. See picture below:

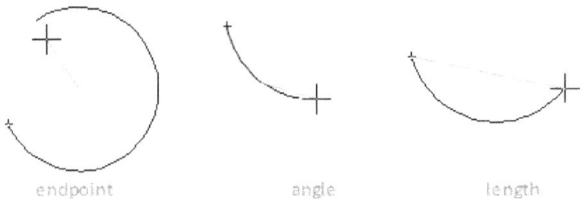

Pic 3.29 Endpoint, angle, and length

- Direction: Specifying tangent for the arc segment.

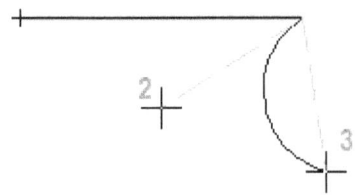

Pic 3.30 Direction

- (2) is the tangent direction from the arc's start point.

- (3) is the last point of the arc. You can use Ctrl to draw counter clockwise.

- Line: Change from the arc drawing to line drawing.

- Radius: Determine the radius from the arc segment.

- Second pt: Determine the second point and the last point from three points the arc.

For a Linetype pattern look at the arguments below:

- PLINEGEN system variable, determine what type of line created in 2 dimensional polyline.

- 0 will create dash at the corner.

- 1 will draw an uninterrupted dotted line.

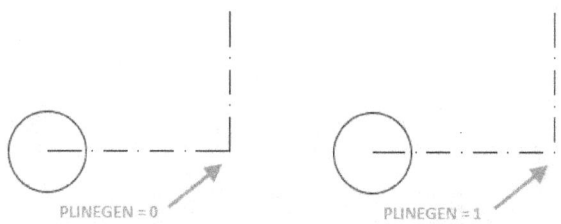

Pic 3.31 The difference between PLINEGEN = 0 and PLINEGEN = 1

See tutorial below to create Polyline:

1. First, create limit from 0,0 to 100,100.

```
Command: limits
Reset Model space limits:
Specify lower left corner or [ON/OFF] <0.0000,0.0000>: 0,0
Specify upper right corner <420.0000,297.0000>: 100,100
```

2. Type polyline and specify the first point to 0,0.

```
Command: POLYLINE
PLINE
Specify start point: 10,10
Current line-width is 0.0000
```

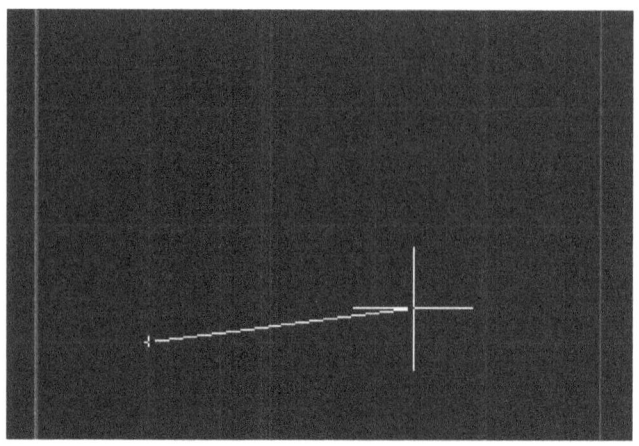

Pic 3.32 Specify the first point of Polyline to 10,10

3. Specify next line to 80,10.

```
Specify next point or [The arc/Halfwidth/Length/Undo/Width]:
<Object Snap Tracking on> 80,10
```

Pic 3.33 Specify second point to 80,10

4. Specify next point to 80,50.

```
Specify next point or [The arc/Close/Halfwidth/Length/Undo/Width]:
80,50
```

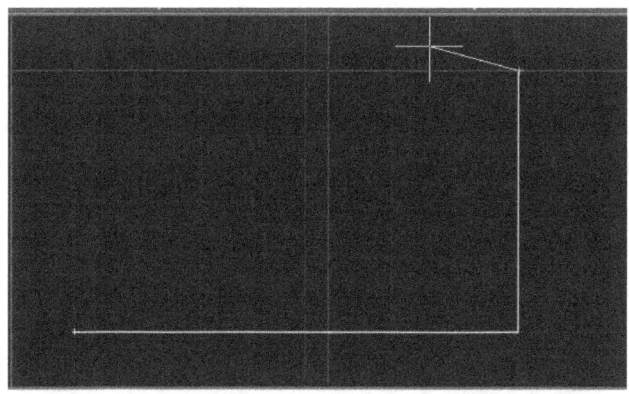

Pic 3.34 Specify next point to,50

5. To create the the arc, insert argument A, and specify angle = 20 and end point of the arc to 10,50.

```
Specify next point or [The arc/Close/Halfwidth/Length/Undo/Width]:
A
Specify endpoint of the arc (hold Ctrl to switch direction) or
[Angle/CEnter/CLose/Direction/Halfwidth/Line/Radius/Second
pt/Undo/Width]: A
Specify included angle: 20
Specify endpoint of the arc (hold Ctrl to switch direction) or
[CEnter/Radius]: 10,50
Specify endpoint of the arc (hold Ctrl to switch direction) or
```

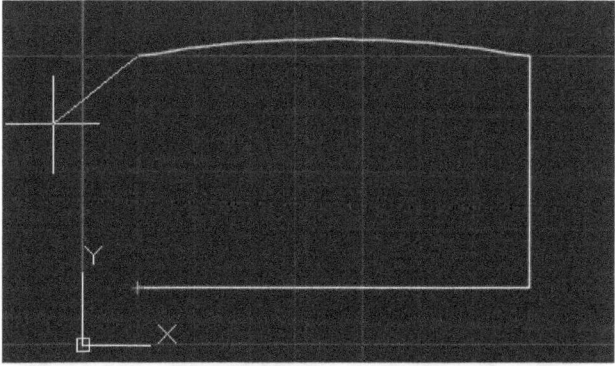

Pic 3.35 The arc creation

6. Back to draw a line, by inserting l argument and type C to close the polyline.

```
Specify endpoint of the arc (hold Ctrl to switch direction) or
[Angle/CEnter/CLose/Direction/Halfwidth/Line/Radius/Second
pt/Undo/Width]: l
```

```
Specify next point or [The arc/Close/Halfwidth/Length/Undo/Width]:
C
```

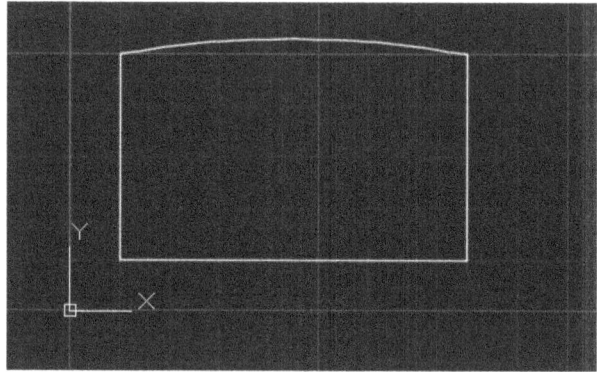

Pic 3.36 Polyline creation

The second tutorial on creating polyline:

1. Insert polyline set start point to 10,10 set halfwidth to 1.

```
Command: POLYLINE PLINE
Specify start point: 10,10
Specify next point or [The arc/Halfwidth/Length/Undo/Width]: h
Specify starting half-width <5.0000>: 1
Specify ending half-width <1.0000>:
```

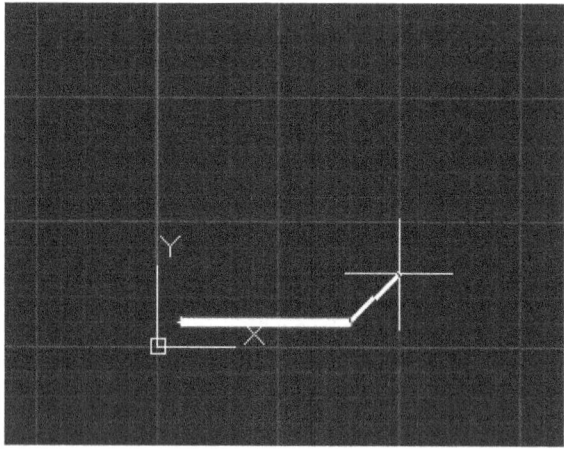

Pic 3.37 Setting halfwidth to 1 and start point to 10,10

2.   Set the next point to 80,10 and 80,50.

```
Specify next point or [The arc/Halfwidth/Length/Undo/Width]: 80,10
Specify next point or [The arc/Close/Halfwidth/Length/Undo/Width]:
80,50
```

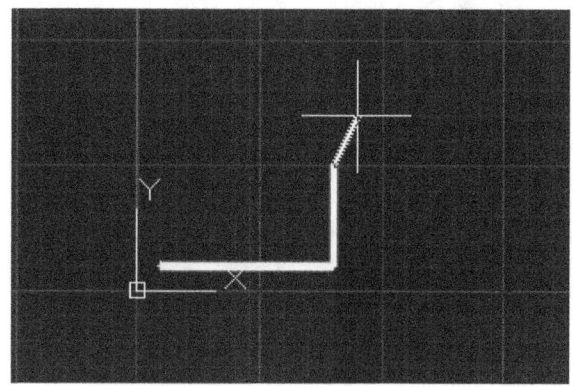

Pic 3.38 Set next point to 80,10 and 80,50

3. Create the arc with radius = 50 and the next point to 10,50.

```
Specify next point or [The arc/Close/Halfwidth/Length/Undo/Width]:
a
Specify endpoint of the arc (hold Ctrl to switch direction) or
[Angle/CEnter/CLose/Direction/Halfwidth/Line/Radius/Second
pt/Undo/Width]: r
Specify radius of the arc: 50
Specify endpoint of the arc (hold Ctrl to switch direction) or
[Angle]: 10,50
```

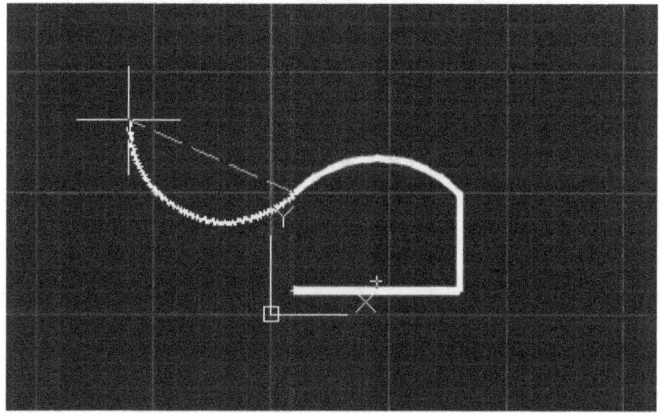

Pic 3.39 Create the arc

4. Choose Close, polyline created with width = 2.

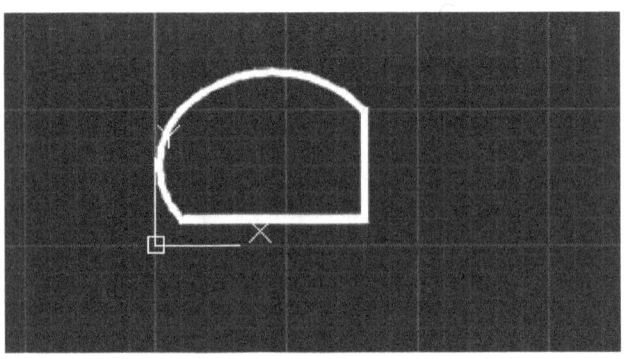

Pic 3.30 Polyline created

3.2.3 Drawing a circle

Circle command used to draw a circle, you can make a circle using some combinations. See examples below to draw a circle in AutoCAD:

1. Type circle, command and let the center to 50,50.

```
Command: CIRCLE
Specify center point for circle or [3P/2P/Ttr (tan tan radius)]:
50,50
```

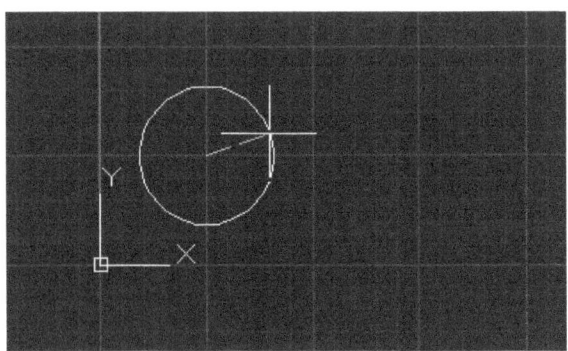

Pic 3.31 Specify center point to 50,50

2. Specify the radius = 50. A circle will be created

```
Specify radius of circle or [Diameter] <50.0000>: 50
```

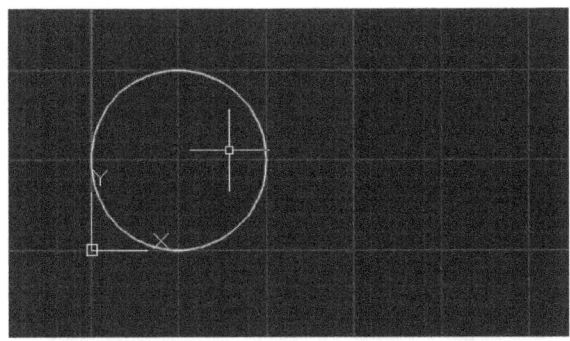

Pic 3.32 Drawing a circle with center = 50,50 and radius = 50

You can also draw a circle by specifying three points. Look at this tutorial:

1. Insert circle command and chooes 3p.

2. Specify the first point to 50,0, the second point to 100,0 and the third point to 50,50.

```
Command: CIRCLE
Specify center point for circle or [3P/2P/Ttr (tan tan radius)]: 3p
Specify the first point on circle: 50,0
Specify second point on circle: 100,0
Specify the third point on circle: 50,50
```

3. AutoCAD will draw a circle based on three points inserted.

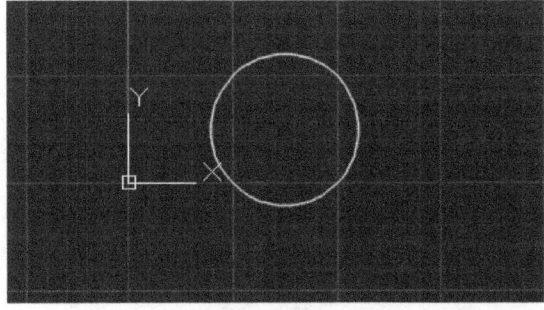

Pic 3.33 Circle created by specifying 3 points

You can also specify 2 points to draw a circle. See steps below:

1. Type "circle" and type 2P for 2 points.

2. Specify the first point 10,10 and 100,10 as second point.

```
Command: CIRCLE
```

```
Specify center point for circle or [3P/2P/Ttr (tan tan radius)]: 2p
Specify first end point of circle's diameter: 10,10
Specify second end point of circle's diameter: 100,10
```

3. If it's saved, you'll see the circle created:

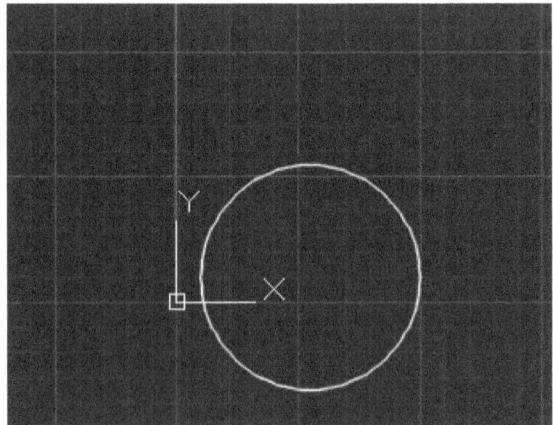

Pic 3.34 Circle created by specifying two points

You can also draw a circle by using 2 tangent and radius. See example below:

1. For example, there are 2 the arcs and I want to draw a circle that tangents to those the arcs and certain radius.

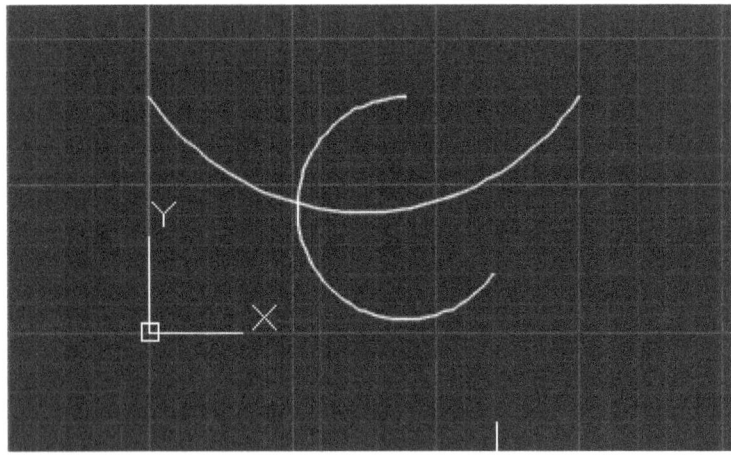

Pic 3.35 Two the arcs

2. Insert circle command and type T in circle parameter.

```
Command: CIRCLE
Specify center point for circle or [3P/2P/Ttr (tan tan radius)]: t
```
3. Click first the arc.

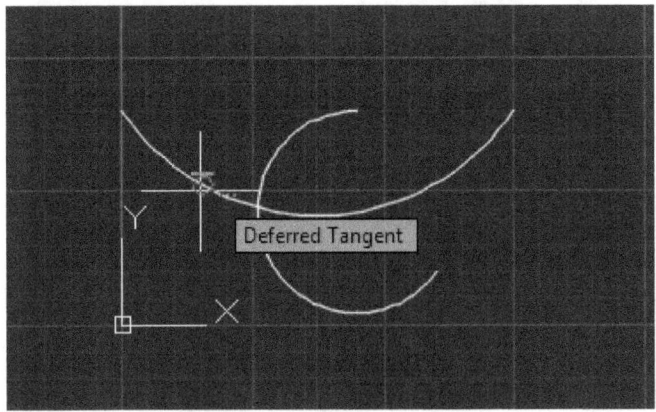

Pic 3.36 Click first the arc

4. Click on the second arc.

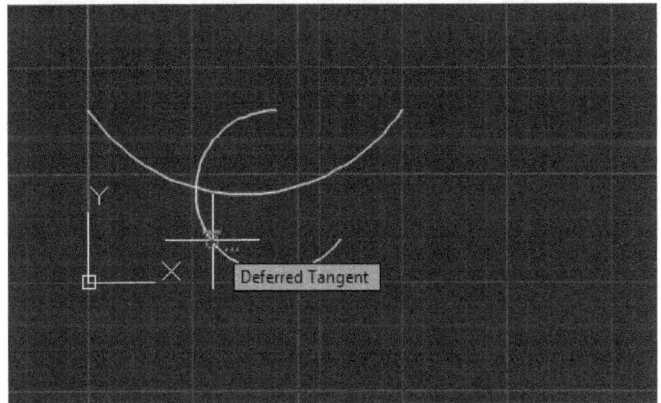

Pic 3.37 Click on second the arc

5. Specify the radius, for this example, I use 50.

```
Command: CIRCLE
Specify center point for circle or [3P/2P/Ttr (tan tan radius)]: t
Specify point on object for first tangent of circle:
Specify point on object for second tangent of circle:
Specify radius of circle <50.0000>: 50
```
6. Click Enter, the circle will be created that tangent to those two the arcs.

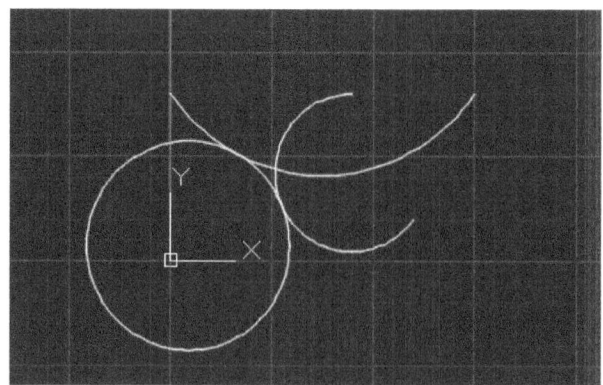

Pic 3.38 Circle with Tan-tan radius

Forth method to draw a circle is by specifying three tangent. See the example below:

1. For example, there are three lines like picture below:

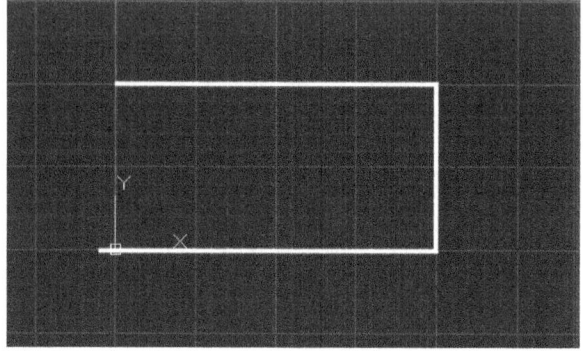

Pic 3.39 Three lines as tangents

2. Click on Circle > Tan, Tan, Tan menu.

Pic 3.40 Menu Circle > Tan, Tan, Tan

3. Click on the first line.

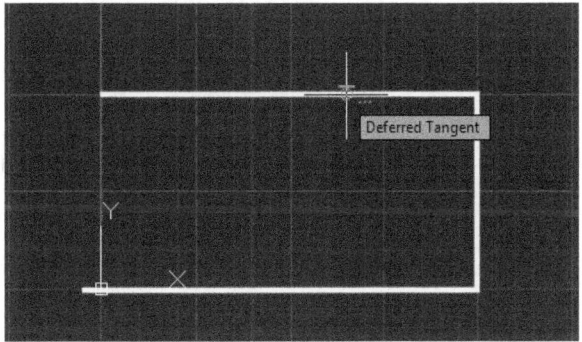

Pic 3.41 Click on the first line

4. Click on the second line.

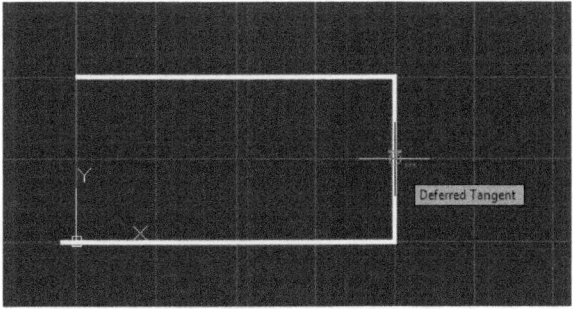

Pic 3.42 Click on second line

5. Click on the third line.

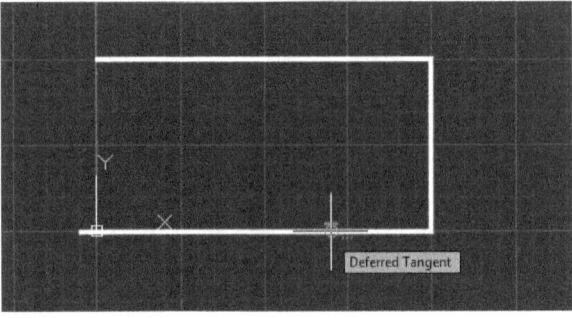

Pic 3.43 Click on the third line

6. The circle will be created.

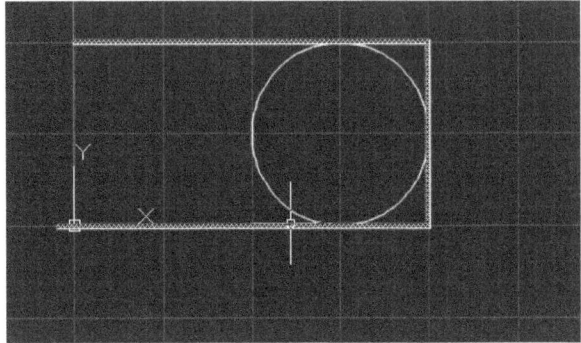

Pic 3.44 The circle created

3.2.4 Drawing The arc

The arc can be created using some methods. First by specifying three points. See tutorial below:

1. Type "the arc" in command line.

2. Specify start point to 0,0.

3. Specify second point to 100,50.

4. Specify the third point to 150,0.

```
Command: THE ARC
Specify start point of the arc or [Center]: 0,0
Specify second point of the arc or [Center/End]: 100,50
Specify end point of the arc: 150,0
```

5. The arc will be created:

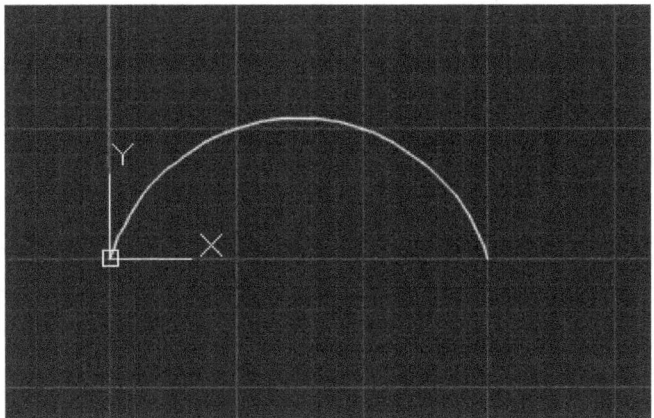

Pic 3.45 The arc created

The second method of creating the arc is by specifying start, center and angle. See steps below:

1. Execute "the arc" command.

2. Specify start point to 0,0.

3. Specify center point of the arc to 50,0.

4. Choose angle and set to -45 degrees.

```
Command: THE ARC
Specify start point of the arc or [Center]: 0,0
Specify second point of the arc or [Center/End]: C
Specify center point of the arc: 50,0
Specify end point of the arc (hold Ctrl to switch direction) or
[Angle/chord Length]: -45
```

5. See picture below for the arc created:

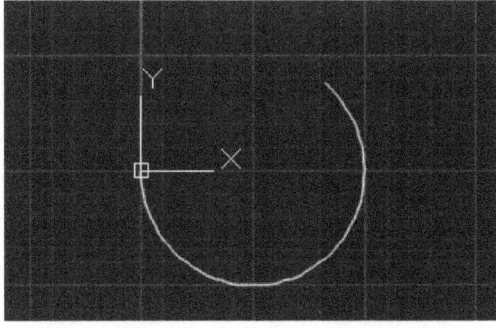

Pic 3.46 The arc created by start point, center, and angle

The second method to create the arc is by specifying Start, Center, Length.

1. Execute the arc command, specify start point of the arc to 0,0.

```
Command: THE ARC
Specify start point of the arc or [Center]: 0,0
Specify second point of the arc or [Center/End]: C
```

2. In Specify second point of the arc, click C to specify Center.

3. Specify center to 50,-40.

4. Insert L to specify length.

5. Specify length = 40.

```
Specify center point of the arc: 50,-40
Specify end point of the arc (hold Ctrl to switch direction) or
[Angle/chord Length]: L
Specify length of chord (hold Ctrl to switch direction): 40
```

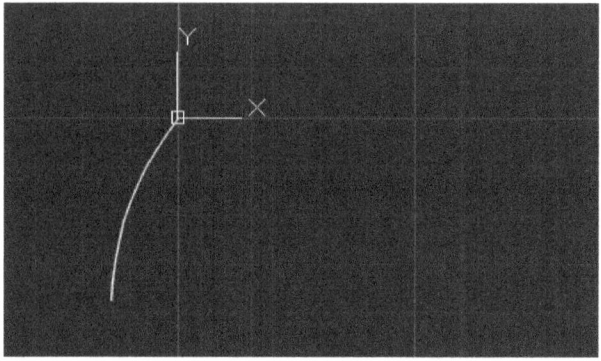

Pic 3.47 The arc created using the arc, center, and length

Next the arc type creation method is by specifying Start End angle. Just enter the start point, end point, and angle. See steps below:

1. Run The arc command, and specify start point to 0,0.

2. Choose E to specify "End point" method.

3. Specify point 100,100 for end point.

```
Command: THE ARC
Specify start point of the arc or [Center]: 0,0
Specify second point of the arc or [Center/End]: E
Specify end point of the arc: 100,100
```

4. Insert A to specify Angle.

5. Type -30 for the angle.

```
Specify center point of the arc (hold Ctrl to switch direction) or
[Angle/Direction/Radius]: A
Specify included angle (hold Ctrl to switch direction): -30
```

6. The arc will be created.

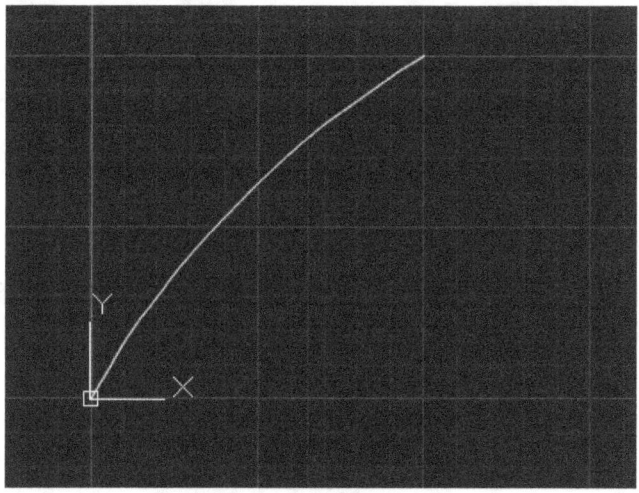

Pic 3.48 The arc created using Start End, and Angle

Next the arc type is Start End, Direction. Here's how to create it:

1. Execute the arc command.

2. Specify start point to 0,0.and type E to select **End** method.

3. Specify End point to 100,0.

```
Command: THE ARC
Specify start point of the arc or [Center]: 0,0
Specify second point of the arc or [Center/End]: E
Specify end point of the arc: 100,0
```

4. Choose D for direction.

```
Specify center point of the arc (hold Ctrl to switch direction) or
[Angle/Direction/Radius]: D
```

5. Specify tangent direction to -45.

```
Specify tangent direction for the start point of the arc (hold Ctrl
to switch direction): -45
```

6. The arc will be created:

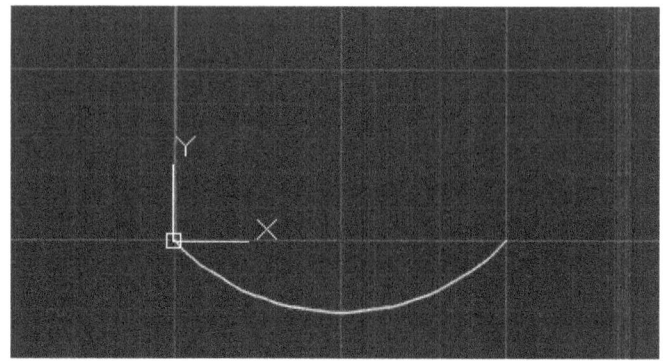

Pic 3.49 The arc created using Start, End, Direction method

Another method to create the arc is by using Start End Radius method. See steps below:

1. Click on The arc > Start, End, Radius.

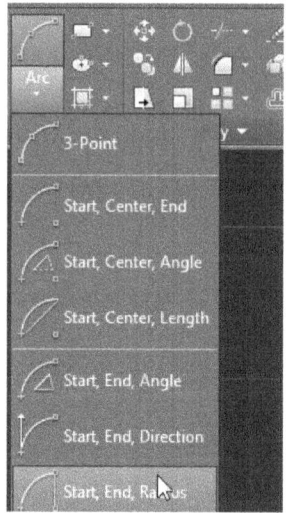

Pic 3.50 Start, End, Radius menu

2. This will set the Start, End, Radius method to create the arc.

3. Specify start point of the arc to 0,0.

```
Command: _the arc
Specify start point of the arc or [Center]: 0,0
Specify second point of the arc or [Center/End]: _e
```

4. Specify end point of the arc to 100,100.

```
Specify end point of the arc: 100,100
```

5. Specify radius to 90.

```
Specify center point of the arc (hold Ctrl to switch direction) or
[Angle/Direction/Radius]: _r
Specify radius of the arc (hold Ctrl to switch direction): 90
```

6. You can see the result below:

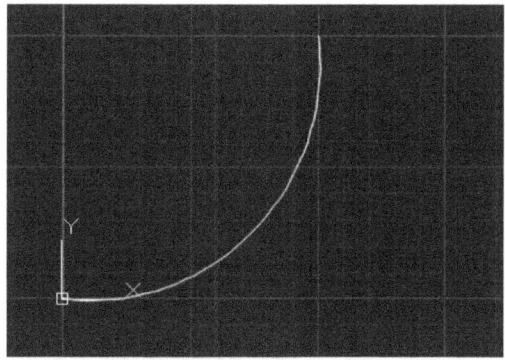

Pic 3.51 The arc created

You can also use Center, Start, End method to create the arc. See steps below:

1. Insert the arc command, then enter C argument for Center.

```
Command: THE ARC
Specify start point of the arc or [Center]: C
```

2. Specify center of the arc = 50,0, and set start point to 0,0.

```
Specify center point of the arc: 50,0
Specify start point of the arc: 0,0
```

3. Specify end point to 100,100.

```
Specify end point of the arc (hold Ctrl to switch direction) or
[Angle/chord Length]: 100,0
```

4. You can see the result as below:

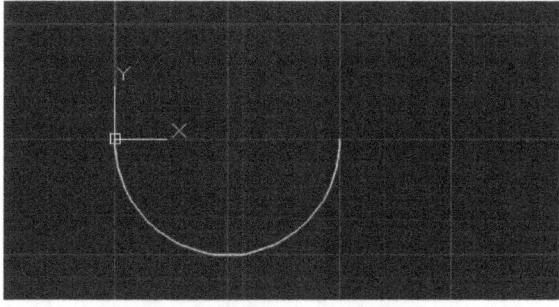

Pic 3.52 Specify center, start, end

Next method is Center Start Angle to create the arc. See steps below:

1. Execute "the arc" command.

2. Choose C for Center.

3. Specify the center point to 50,0.

```
Command: THE ARC
Specify start point of the arc or [Center]: C
Specify center point of the arc: 50,0
```

4. Specify the start point of the arc to 0,0, then choose A for Angle.

```
Specify start point of the arc: 0,0
Specify end point of the arc (hold Ctrl to switch direction) or
[Angle/chord Length]: A
```

5. Specify the angle = 45 degrees.

```
Specify included angle (hold Ctrl to switch direction): 45
```

6. The result will be as below:

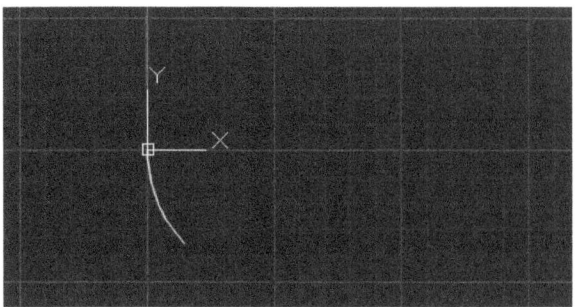

Pic 3.53 The arc creating using "center, start, angle"

Next method is by using Center, Start, and Length. See the steps below:

1. Run "the arc" command.

2. Choose C for Center.

```
Command: THE ARC
Specify start point of the arc or [Center]: C
```

3. Specify the center point of the arc to 50,0.

4. Specify the start point to 0,0.

5. Choose L in Angle/Chord Length.

```
Specify center point of the arc: 50,0
Specify start point of the arc: 0,0
Specify end point of the arc (hold Ctrl to switch direction) or
[Angle/chord Length]: L
```

6. Specify the length of the arc to 100.

```
Specify length of chord (hold Ctrl to switch direction): 100
```

7. The arc will be as below:

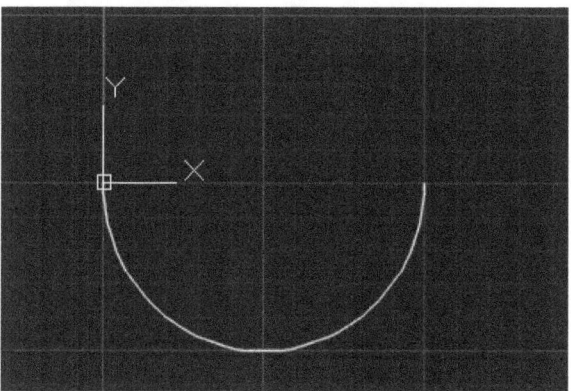

Pic 3.54 The arc already created

To draw a new the arc connected from existing the arc, you can use **Continue**. Here are the steps:

1. After creating the arc.

2. Click on **arc > Continue** in the **Home > Draw** ribbon.

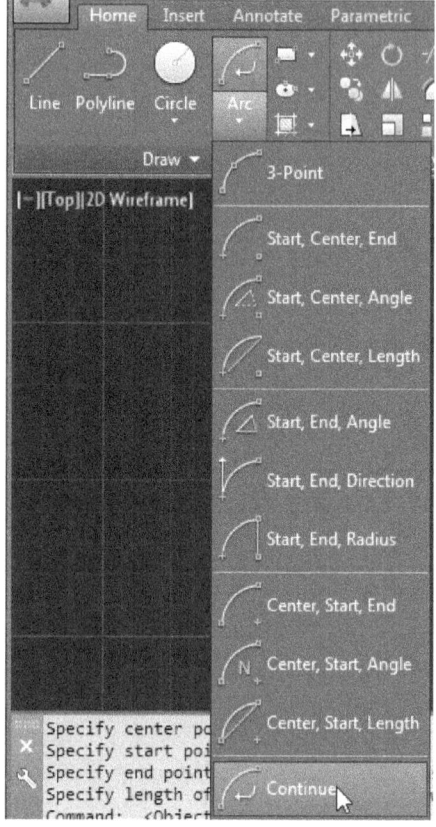

Pic 3.55 Arc > Continue menu

3. You can continue creating the arc from the existing arc.

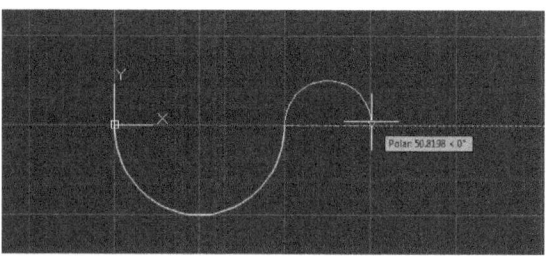

Pic 3.56 Continue creating the arc from the existing arc

4. If you click Enter, another segment of the arc will created.

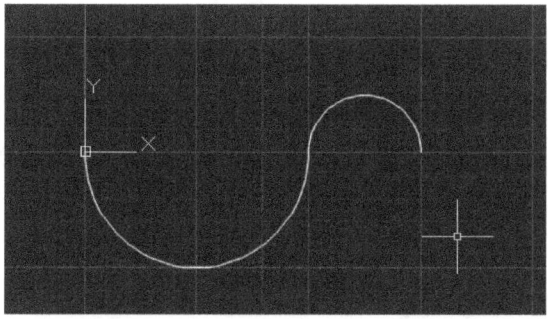

Pic 3.57 The arc segment created

5. By iterating steps above, you can create the arc as much as you want.

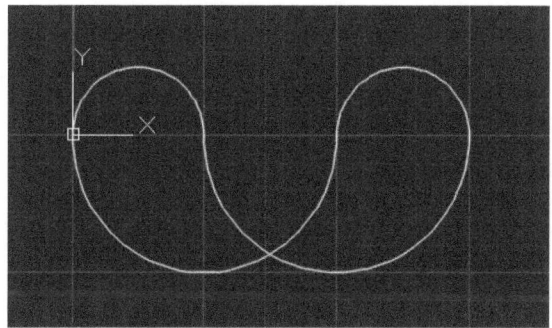

Pic 3.58 The arcs created

3.2.5 Drawing Rectangle

To draw rectangle, AutoCAD, use "rectang" command. This will automatically create polyline in rectangle. Just like the arc, there are more than one methods to create rectangle.

Rectang function has many arguments. You can see the example on commands below:

```
Current settings: Rotation = 0
Specify first corner point or
[Chamfer/Elevation/Fillet/Thickness/Width]:
```

Notes on the arguments:

- First Corner Point: Specifying the first point of rectangle.

- Other Corner Point: Specifying the other point of rectangle.

- Area: Creating rectangle using area, length and width. If Chamfer or Fillet option active, the effect will be appeared on the corner of rectangle.

- Dimensions: Creating rectangle by entering the length and width.

- Rotation: Creating rectangle using certain angle rotation.

- Chamfer: Setting the chamfer of rectangle.

- Elevation: Setting the elevation of rectangle.

- Fillet: Setting fillet radius of rectangle.

- Thickness: Setting the thickness of rectangle.

- Width: Setting the line's width of rectangle.

First tutorial describes how to create simple rectangle, without fillet and width. See steps below:

1. Run "rectang" command.

2. Specify the first corner to 0,0.

```
Command: RECTANG
Specify first corner point or
[Chamfer/Elevation/Fillet/Thickness/Width]: 0,0
```

3. Specify other corner to 75,0.

```
Specify other corner point or [Area/Dimensions/Rotation]: 75,50
```

4. A simple rectangle will be created on the drawing area.

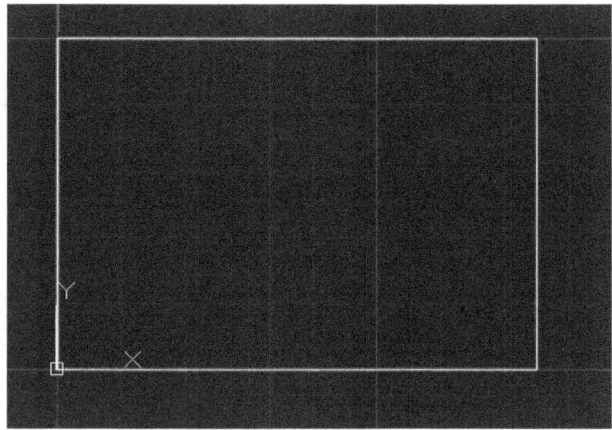

Pic 3.59 Rectangle created using RECTANG command

The rectangle may has a chamfer. See steps below to create the rectangle with a chamfer.

1. Run "RECTANG" command.

2. Choose C on "Specify first corner" to activate the chamfer.

```
Command: RECTANG
Specify first corner point or
[Chamfer/Elevation/Fillet/Thickness/Width]: C
```

3. On Specify first chamfer distance, set to 3, on specify second chamfer distance, set to 3.

```
Specify first chamfer distance for rectangles <0.0000>: 3
Specify second chamfer distance for rectangles <3.0000>: 3
```

4. Specify first corner point = 0,0. Then Specify second corner point to 50,50.

```
Specify first corner point or
[Chamfer/Elevation/Fillet/Thickness/Width]: 0,0
Specify other corner point or [Area/Dimensions/Rotation]: 50,50
```

5. The result is a rectangle with a chamfer on the corner. See picture below:

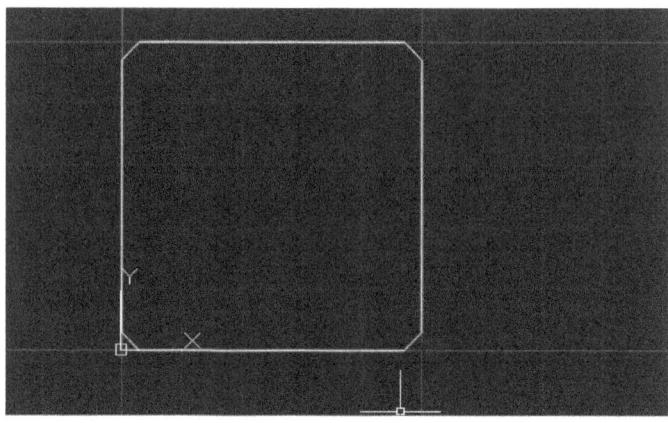

Pic 3.60 a rectangle with chamfer

You can also create fillets on the corner of rectangle. See steps below:

1. Run "rectang" command.

```
Command: RECTANG
```

2. On "Specify first corner", click F.

3. Specify the fillet radius to 3.

4. Then specify first corner point = 0,0.

5. Then Specify another corner point = 50,50.

```
Specify first corner point or
[Chamfer/Elevation/Fillet/Thickness/Width]: F
Specify fillet radius for rectangles <3.0000>: 3
Specify first corner point or
[Chamfer/Elevation/Fillet/Thickness/Width]: 0,0
Specify other corner point or [Area/Dimensions/Rotation]: 50,50
```

6. The result is a rectangle with fillet:

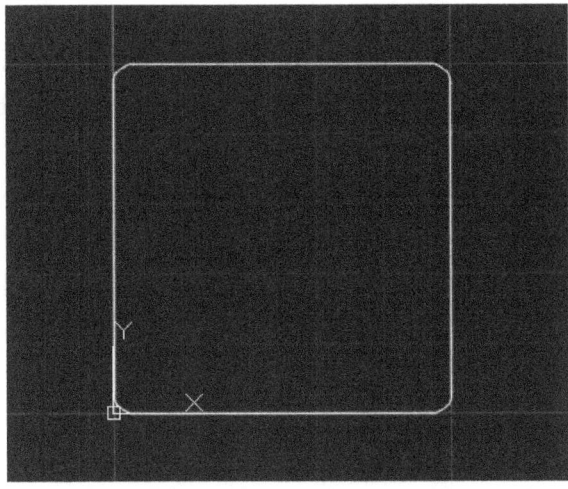

Pic 3.61 A rectangle with fillet

You can also change the width argument of a rectangle to draw a rectangle with custom width. The steps are:

1. Run "RECTANG" command.

```
Command: RECTANG
Current rectangle modes:   Fillet=3.0000
```

2. On "Chamfer/Elevation/Fillet", click w.

3. Set the width to 1.

```
Specify first corner point or
[Chamfer/Elevation/Fillet/Thickness/Width]: w
Specify line width for rectangles <0.0000>: 1
```

4. Specify the first corner point = 0,0. And then the second corner point =100,50.

```
Specify first corner point or
[Chamfer/Elevation/Fillet/Thickness/Width]: 0,0
Specify other corner point or [Area/Dimensions/Rotation]: 100,50
```

5. Rectangle created will have custom width.

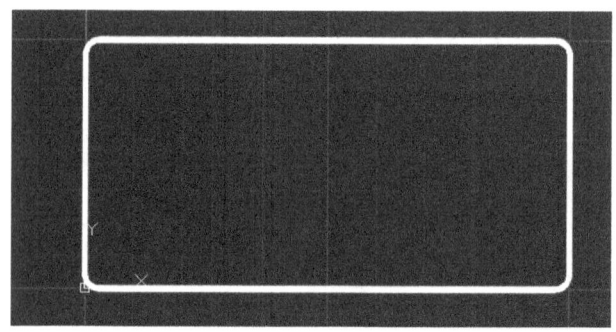

Pic 3.62 Create Rectangle

✓ *Exercise Drawing Basic Shapes and Edit Sketches*

For this AutoCAD tutorial, type in "Rectangle" and press Enter to initiate the command. Start at the CenterPoint and end at 10/50.

Start a circle at 0/47.5 and confirm by pressing enter. Set the radius to 8. If you made a mistake, simply double-click on the sketch you want to edit. In the popped-up window edit the values.

Start a center ellipse at 0/30. Set the major radius parallel to the X-Axis to 70 and set the minor radius to 30.

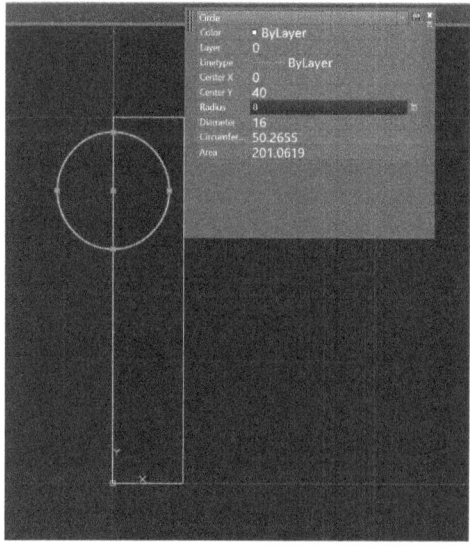

✓ *Exercise Drawing a Second Circle with Object Snap Enabled*

Draw a second circle at 25/47.5. Turn on Object Snap with by pressing F3 and guide the radius of the circle parallel to the Y-Axis until you intersect with the ellipse. Click when you see a green Cross. Draw a line starting at 10/55, you might want to turn off Object Snap, so the starting point will not get caught at the corner of the rectangle. When you have placed the starting point turn, on Object Snap with the "Tangent" option enabled. Draw a line at a 65° angle until it snaps with the second circle. Start a second line at the top right corner of the rectangle. Enable "Nearest" in Object Snap option draw a line in a 130° angle, snapping to the first circle.

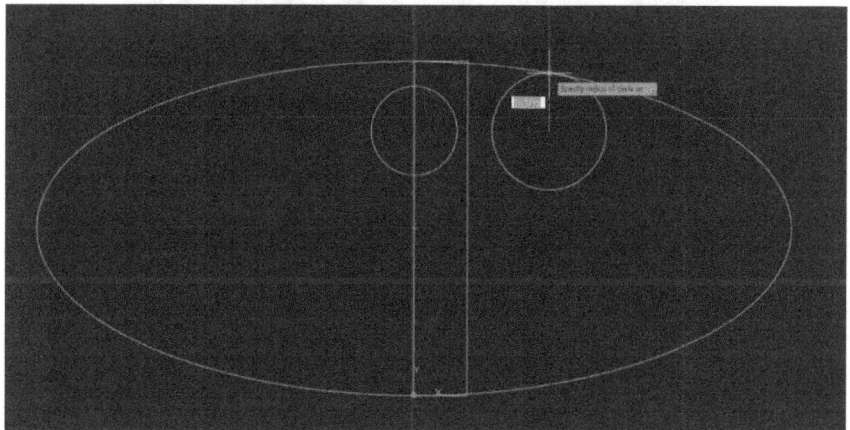

3.2.6 Drawing Polygon

Polygon will draw a polygon with a custom number of sides. The default number of sides is 4, but you can customize. A polygon can be inscribed or circumscribed. See tutorial below to draw polygon:

1. First, draw a circle

2. Specify the center of the circle to 25.25.

3. Specify the circle radius to 25.

```
Command: CIRCLE
Specify center point for circle or [3P/2P/Ttr (tan tan radius)]:
25,25
Specify radius of circle or [Diameter]: 25
```

4. A circle will be created with center = 25.25 and radius = 25. Create the circle to help you to distinguish between inner or outer polygon.

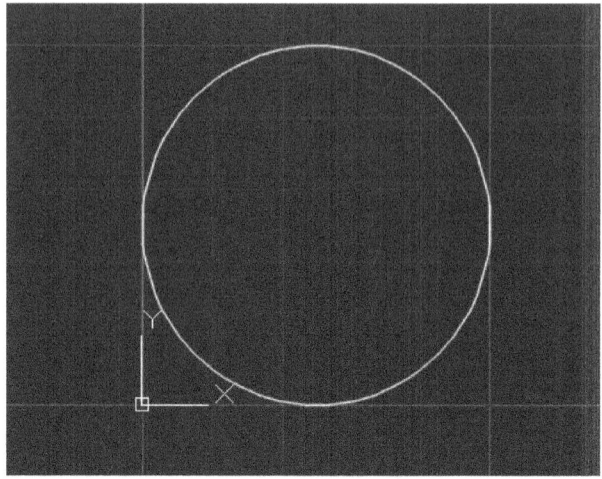

Pic 3.63 Circle created

5. Execute "polygon" command.

6. Enter the number of sides to 5.

```
Command: POLYGON
Enter number of sides <4>: 5
```

7. Specify the center of polygon to 25,25.

8. For first polygon, I choose the inscribed by typing I

```
Specify center of polygon or [Edge]: 25,25
Enter an option [Inscribed in circle/Circumscribed about circle]
<I>: i
```

9. Specify the radius to 25.

```
Specify radius of circle: 25
```

10. You can see the polygon created, but inscribed in the circle.

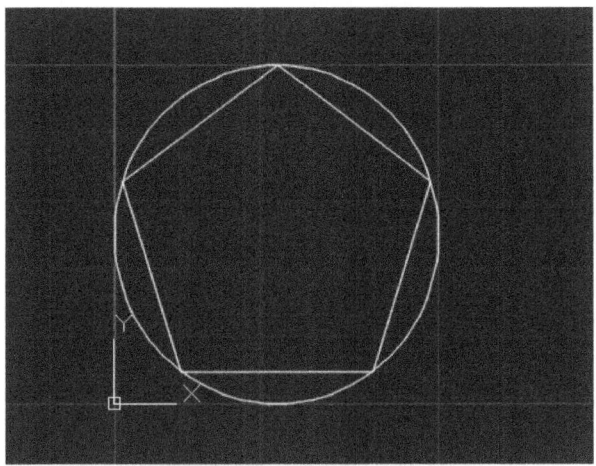

Pic 3.64 Polygon inserted inside the circle

If you want to create circumscribed polygon. Use steps below:

1. Type "polygon" in the command prompt

2. Specify the center of to 25.25.

```
Command: POLYGON
Enter number of sides <5>:
Specify center of polygon or [Edge]: 25,25
```

3. Type C to specify circumscribed.

```
Enter an option [Inscribed in circle/Circumscribed about circle]
<I>: C
```

4. Define radius of circle = 25.

```
Specify radius of circle: 25
```

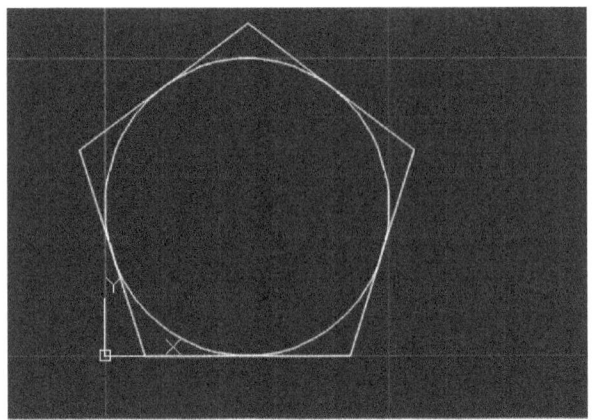

Pic 3.65 The circumscribed polygon inserted

5. You can compare the inscribed polygon yang di dalam atau di luar lingkaran seperti berikut ini:

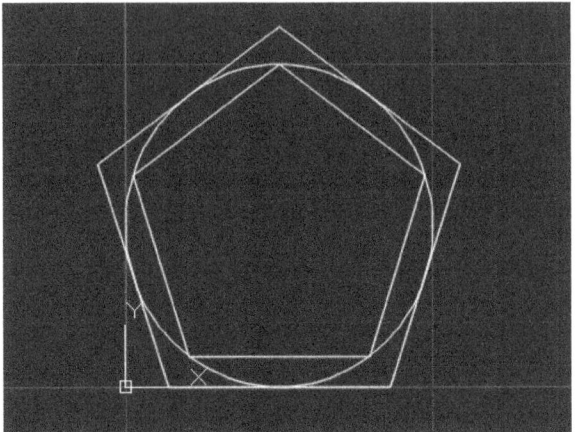

Pic 3.66 Polygons created

3.2.7 Drawing Ellipse

To draw ellipse, you have to define the long axis and short axis, see picture below for the details

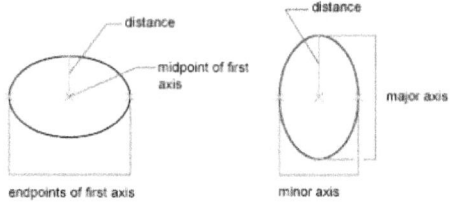

Pic 3.67 Ellipse drawing

Below are steps to create ellipse:

1. Type "ellipse" to create ellipse.

2. Specify axis' end point to 100,50.

3. Specify other endpoint of axis to 0,50.

4. Specify the distance to 20.

```
Command: ELLIPSE
Specify axis endpoint of ellipse or [The arc/Center]: 100,50
Specify other endpoint of axis: 0,50
Specify distance to other axis or [Rotation]: 20
```

5. The ellipse will be created.

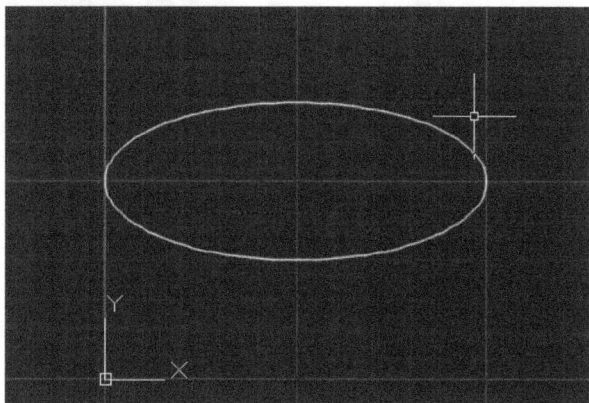

Pic 3.68 Ellipse created

You can create the arc from ellipse. See steps below:

1. Execute "ellipse" command.

2. Choose A to specify the arc.

```
Command: ELLIPSE
Specify axis endpoint of ellipse or [The arc/Center]: A
```

3. Specify axis' end point to 100,0.

4. Specify axis' other end point to 0,0.

```
Specify axis endpoint of elliptical the arc or [Center]: 100,0
Specify other endpoint of axis: 0,0
```

5. Specify distance to other axis = 30 to create the ellipse.

```
Specify distance to other axis or [Rotation]: 30
```

6. Specify start angle to 50 and end angle to 10.

```
Specify start angle or [Parameter]: 50
Specify end angle or [Parameter/Included angle]: 10
```

7. Ellipse the arc will be created.

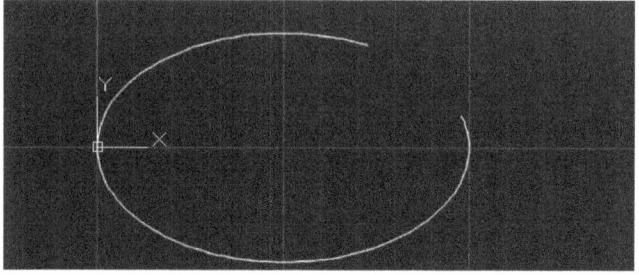

Pic 3.69 Ellipse the arc created

You can also create rotated ellipse, see steps below:

1. Type "ellipse" in command line.

2. Specify ellipse's end point to 100,0.

3. Specify ellipse's other end point to 0,50.

```
Command: ELLIPSE
Specify axis endpoint of ellipse or [The arc/Center]: 100,0
Specify other endpoint of axis: 0,50
```

4. Specify distance to 45.

```
Specify distance to other axis or [Rotation]: 45
```

5. Specify rotation to 45.

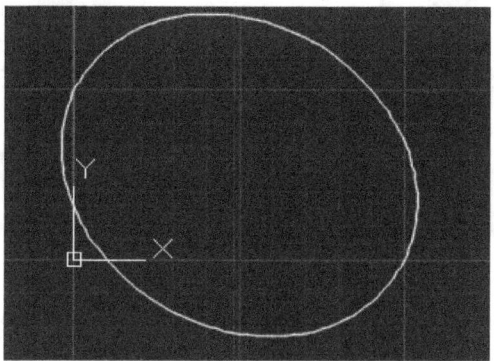

Pic 3.70 Rotated the arc created

3.2.8 Drawing Hatch

Certain area can be hatched, you can also define the hatch type, see example below for drawing hatch:

1. Create two objects. One circle and one polygon with number of sides = 5.

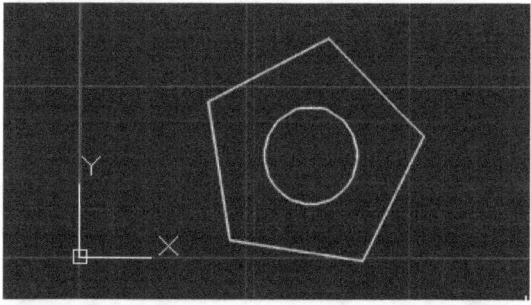

Pic 3.71 Creating two objects, circle and polygon

2. Select both objects by your mouse.

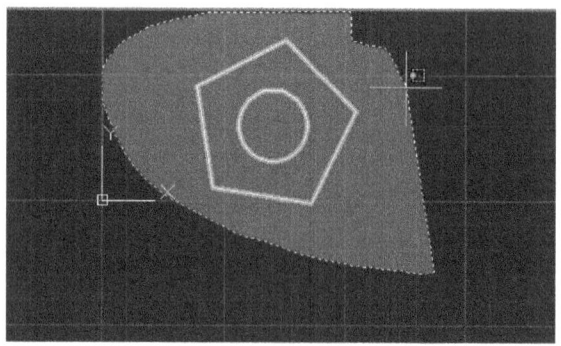

Pic 3.72 Selecting objects

3. Both objects selected, see following pic:

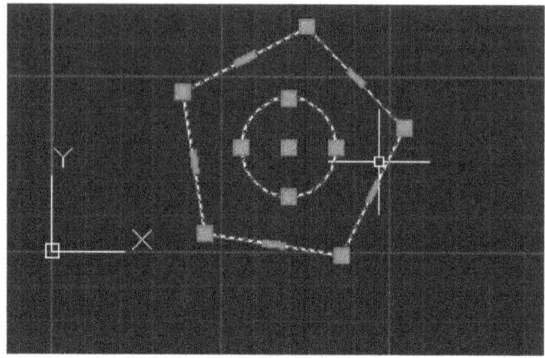

Pic 3.73 Both objects selected

4. Right click until the context menu appears, and select **Group** > **Group**.

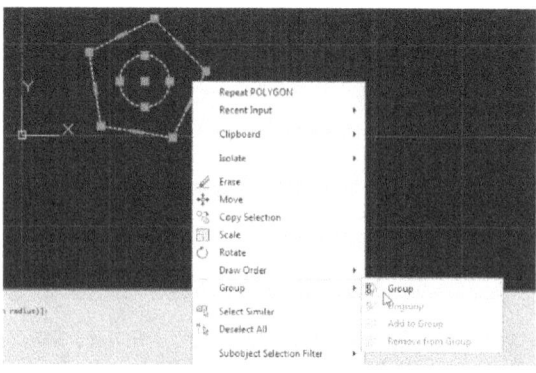

Pic 3.74 Choosing Group > Group menu

5. Objects will be grouped, then choose the object.

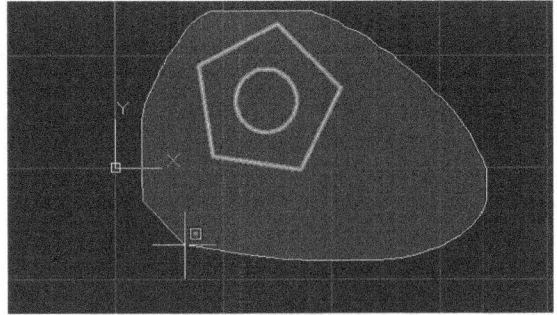

Pic 3.75 Select the grouped object

6. After selected, you can see the objects become one entity (because it's already grouped using Group > Group menu).

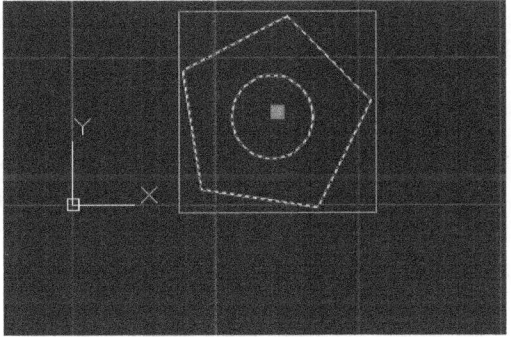

Pic 3.76 Grouped object

7. To hatch area between circle and polygon, click **Hatch** in **Home > Draw** box.

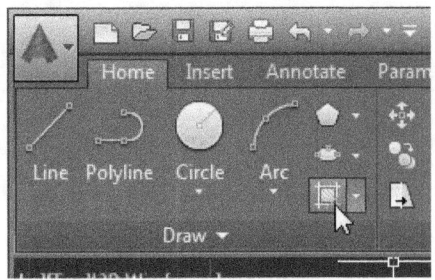

Pic 3.77 Click on Hatch button

91

8. In Pattern box, click on the arrow button to display more patterns.

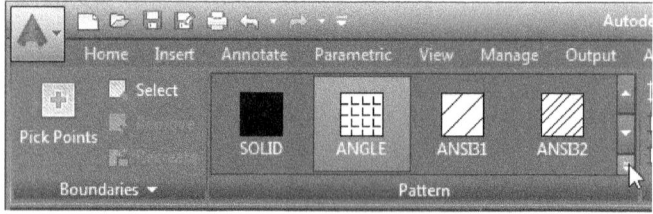

Pic 3.78 Click on arrow to display more patterns

9. You can see list of hatch's pattern.

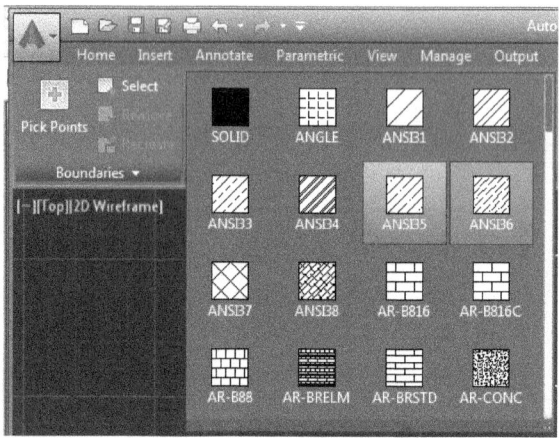

Pic 3.79 Hatch patterns

10. After selecting the pattern, click on the area.

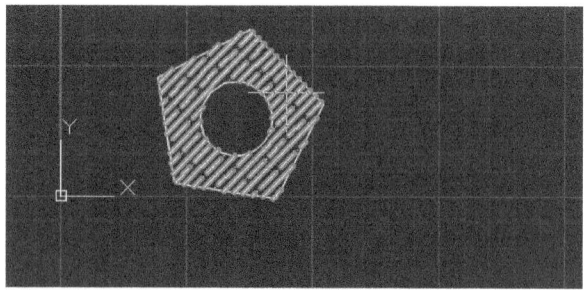

Pic 3.80 Click on the area to be hatched

11. The hatch will be created:

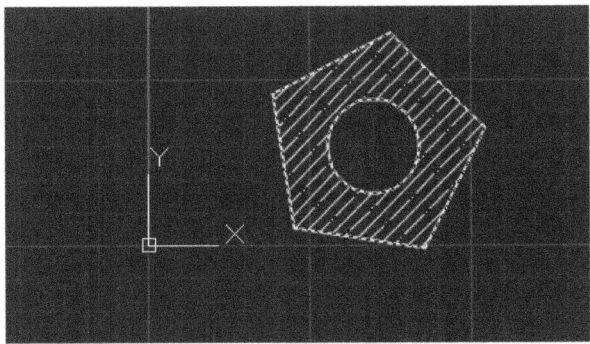

Pic 3.81 Hatched area

3.2.9 Drawing Spline

Spline command is used to create curvy line. You can use Fit method or CV method. See tutorial below:

1. Run "spline" command in AutoCAD.

```
Command: SPLINE
Current settings: Method=Fit    Knots=Chord
```

2. Specify the first point to 0,0 and the next point to 25,25.

```
Specify the first point or [Method/Knots/Object]: 0,0
Enter next point or [start Tangency/toLerance]: 25,25
```

3. Specify next point to 50,0 and 75,25

```
Enter next point or [end Tangency/toLerance/Undo]: 50,0
Enter next point or [end Tangency/toLerance/Undo/Close]: 75,25
```

4. Specify next point to 100,0 and 50,-50. Then click C on your keyboard to close spline.

```
Enter next point or [end Tangency/toLerance/Undo/Close]: 100,0
Enter next point or [end Tangency/toLerance/Undo/Close]: 50,-50
Enter next point or [end Tangency/toLerance/Undo/Close]: C
```

5. See following picture to see the spline result:

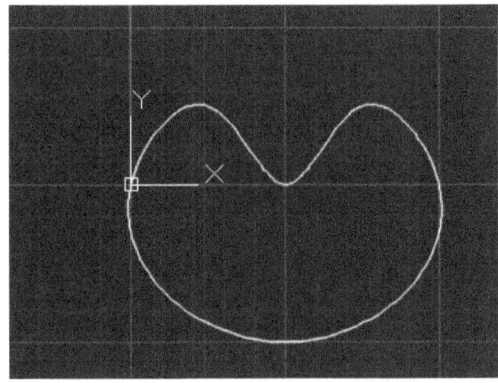

Pic 3.82 Spline result

Spline can also use CV method. See steps below:

1. Enter "spline" command and click "m" on your keyboard to specify Method.

```
Command: SPLINE
Specify the first point or [Method/Degree/Object]: m
```

2. Insert cv to choose cv method for spline creation.

```
Enter spline creation method [Fit/CV] <CV>: cv
Current settings: Method=CV   Degree=3
```

3. Specify the first point to 0,0 and the next point to 25,25.

```
Specify the first point or [Method/Degree/Object]: 0,0
Enter next point: 25,25
```

4. Specify next point to 50,0 and 75,25.

```
Enter next point or [Undo]: 50,0
Enter next point or [Close/Undo]: 75,25
```

5. Insert next point 100,0 and click C butotn on your keyboard to close the spline.

```
Enter next point or [Close/Undo]: 100,0
Enter next point or [Close/Undo]: C
```

6. See following pic for the result.

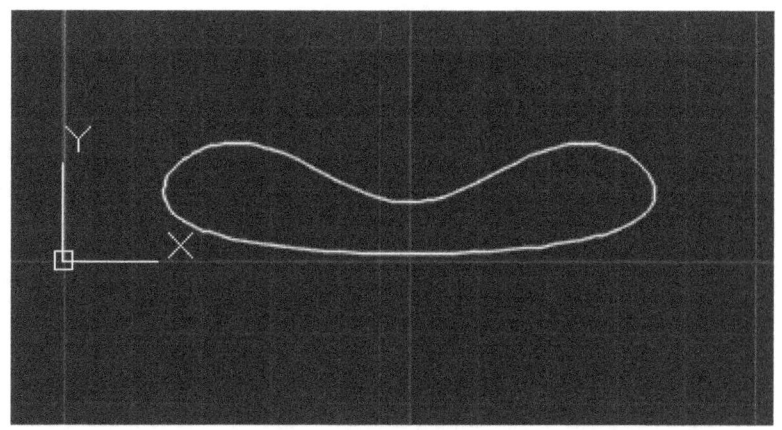

Pic 3.83 Spline result created with cv method

✓ ***Exercise Drawing with The Spline Command***

Continue our previous exercise in 3.2.5, create a Spline starting at the center point. With the Spline tool, you can create a continuous curved sine connecting points. First, you enter the distance, followed by the angle. If you made a mistake type in "U" and press Enter to undo the last step. Enter the following polar coordinates: 20/30°, 5/300°, 5/55°, 10/30°, 5/320°. End with a 230° Angle on the Ellipse. Now type in a "T" to End Tangency and type in 190° for the angle and press Enter.

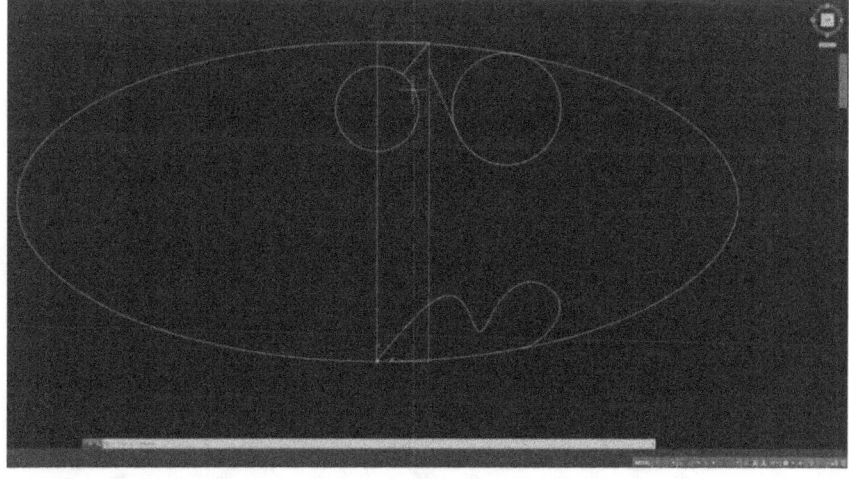

3.2.10 Drawing XLINE

Xline is infinite line, commonly used in construction drawing. Xline command enables you to create infinite line just by specifying two points.

See tutorial below to draw XLINE:

1. Execute "xline" command.

2. Specify the first point to 50,0.

3. Specify second point to 50,10.

```
Command: XLINE
Specify a point or [Hor/Ver/Ang/Bisect/Offset]: 50,0
Specify through point: 50,10
```

4. An infinite vertical line that passes two points specified will be created.

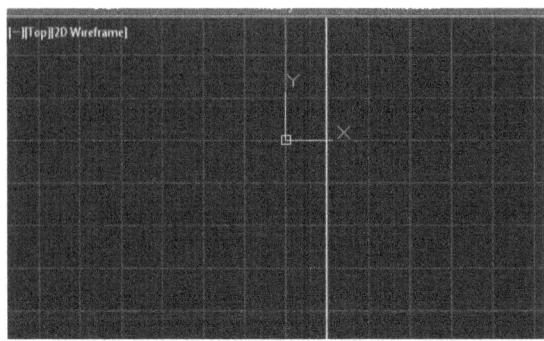

Pic 3.84 Infinite vertical line created with xline

5. To create infinite horizontal line, type "xline".

6. Set the first point to 50,50 and the second point to 100,50.

```
Command: XLINE
Specify a point or [Hor/Ver/Ang/Bisect/Offset]: 50,50
Specify through point: 100,50
```

7. A horizontal xline will be created that passes the two points.

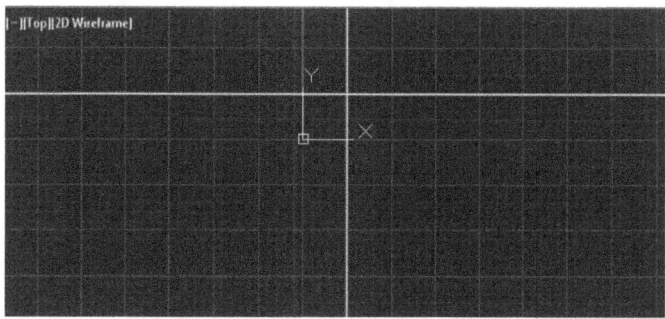

Pic 3.85 Horizontal xline created

8. To create xline with specified angle, first execute "xline".

9. Choose "a" for Angle by clicking "A" button on your keyboard.

10.   Set angle to 30 degrees.

11.   Specify point to 50,50.

```
Command: XLINE
Specify a point or [Hor/Ver/Ang/Bisect/Offset]: a
Enter angle of xline (0) or [Reference]:  30
Specify through point: 50,50
```

12.   See following pic for the result.

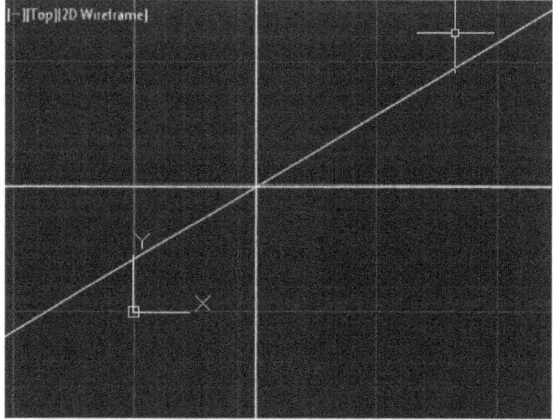

Pic 3.86 Xline with angle

3.2.11 Drawing RAY

Ray similar with xline, but ray have start point. See tutorial below for creating ray line:

1. Type "ray" for Ray line.

2. Specify start point to 50,50.

3. Specify the through point to 75,75 and 100,50 and 100,25.

```
Command: _ray Specify start point: 50,50
Specify through point: 75,75
Specify through point: 100,50
Specify through point: 100,25
```

4. The result will be as below:

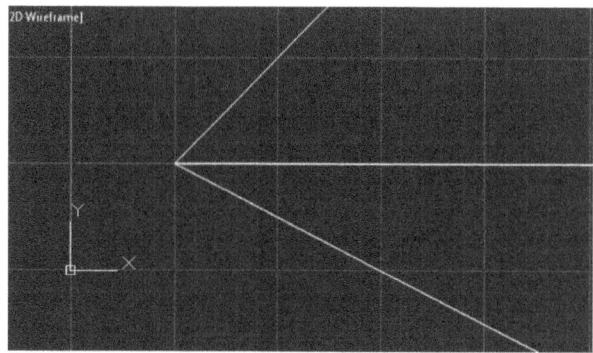

Pic 3.87 Creating line with Ray command

3.2.12 Divide

DIVIDE command divides line or object to some segments. This is suitable for creating dimension's annotation . See tutorial below for the detail:

1. For example there is a line I want to divide into segments.

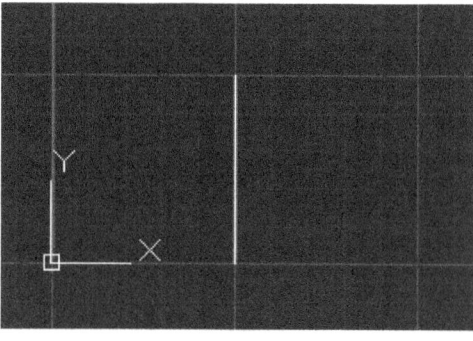

Pic 3.88 A line to divide

2. Type "divide" or click Divide button in **Home > Draw**.

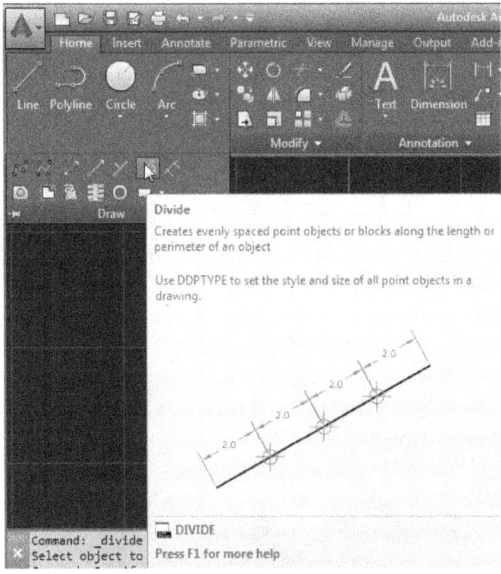

Pic 3.89 Click Divide button

3. Select this line.

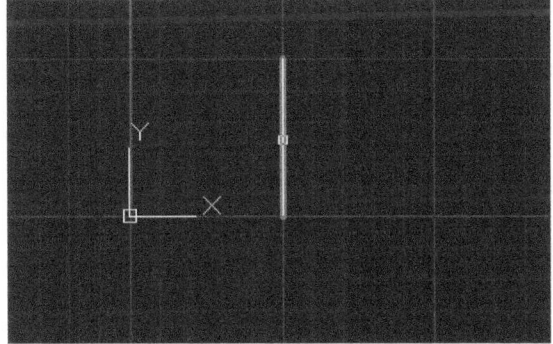

Pic 3.90 Click on the line

4. Select the line you want to divide, selected line will become dotted line.

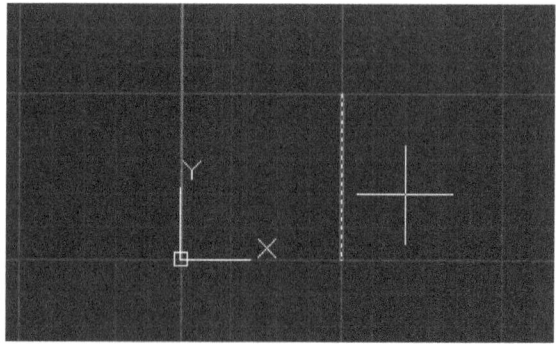

Pic 3.91 Selected Line become dotted line

5. Enter number of segment to 5, this will create 5 segments or 4 points inside the line.

```
Command: _divide
Select object to divide:
Enter the number of segments or [Block]: 5
```

6. If the line moved, you can see 4 points, the points were created by "divide" command.

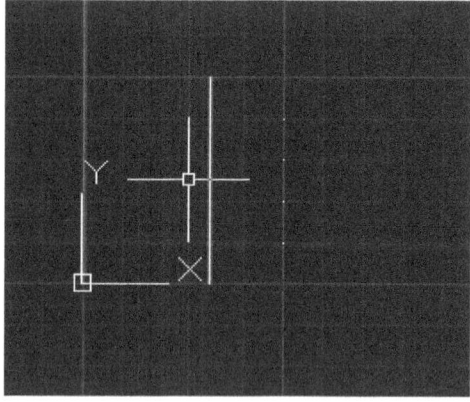

Pic 3.92 Four points that divide line to 5 segments already created

3.2.13 Drawing Helix

Helix command creates helix object. You just have to specify the lower diameter, upper diameter and the height. See example below:

1. Type "helix" in the command prompt.

```
Command: HELIX
Number of turns = 3.0000    Twist=CCW
```

2. Specify center point of the base to 50,50. Then specify base's radius to 30 and top's radius to 30. In this example, I use same radius for top and base.

3. Specify the height to 50.

```
Specify center point of base: 50,50
Specify base radius or [Diameter] <22.3607>: 30
Specify top radius or [Diameter] <30.0000>: 30
Specify helix height or [Axis endpoint/Turns/turn Height/tWist]
<50.9902>: 50
```

4. The helix will be created.

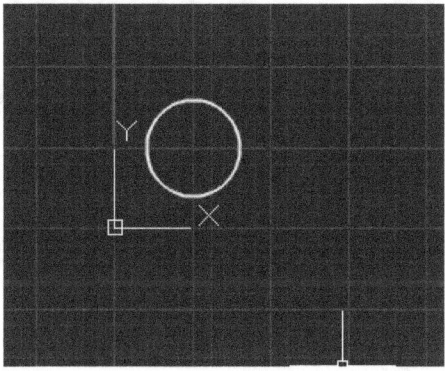

Pic 3.93 Helix created

5. What you see is circle because helix is 3d object, and you only see from the top. To view from side, change the WCS navigator.

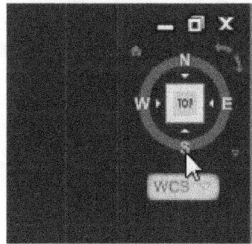

Pic 3.94 Changing the WCS navigator

6. Change the view to front.

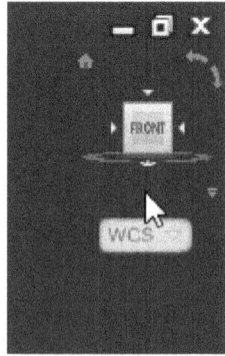

Pic 3.95 Changing WCS view to front

7. See following pic for the helix seen from side view.

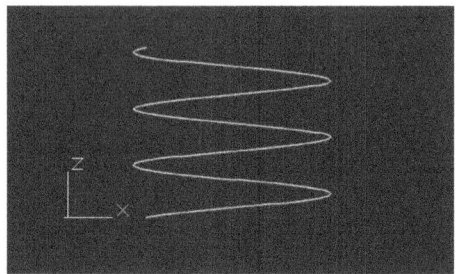

Pic 3.96 Helix seen in side view

8. Back again to TOP view.

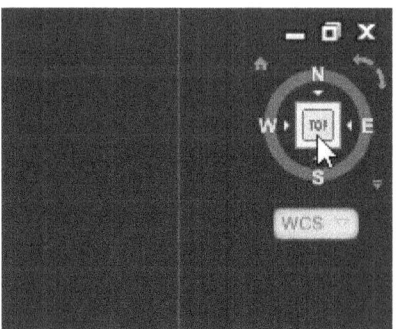

Pic 3.97 Returning to TOP view

9. The helix will be circle again, because the base radius = top radius.

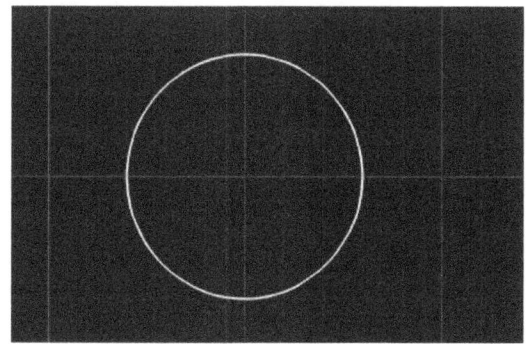

Pic 3.98 Helix seen from above

3.2.14 Drawing Donut

Donut command used to create object similar to donut, that is a circle with inside diameter, and outside diameter. See following example:

1. Type "donut" command.

2. Set inside diameter to 50 and outside diameter to 70.

```
Command: DONUT
Specify inside diameter of donut <50.0000>: 50
Specify outside diameter of donut <70.0000>: 70
```

3. Specify center coordinate to 50,50.

```
Specify center of donut or <exit>: 50,50
```

4. See the result in the following pic.

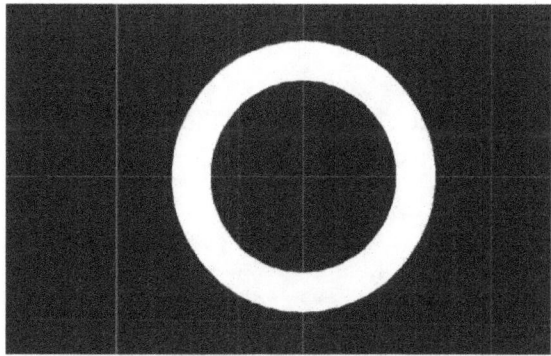

Pic 3.99 Donut created

3.3   Modify 2D Drawing

The 2D Drawing already created, can be modified again. AutoCAD has lots of functions to accommodate modification.

3.3.1 Move

Move command used to move existing object to other place. It's common to use relative coordinate or polar coordinate to move the object. See example below:

1. For example, I have object like this.

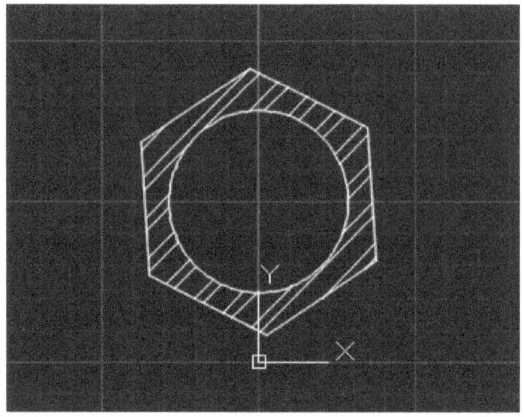

Pic 3.200 Object that will be moved

2. Type "move", then choose object you want to move.

```
Command: MOVE
Select objects: 3 found, 1 group
Select objects: click object
```

3. Click on the object, and select the base point by inserting coordinate or click using your mouse.

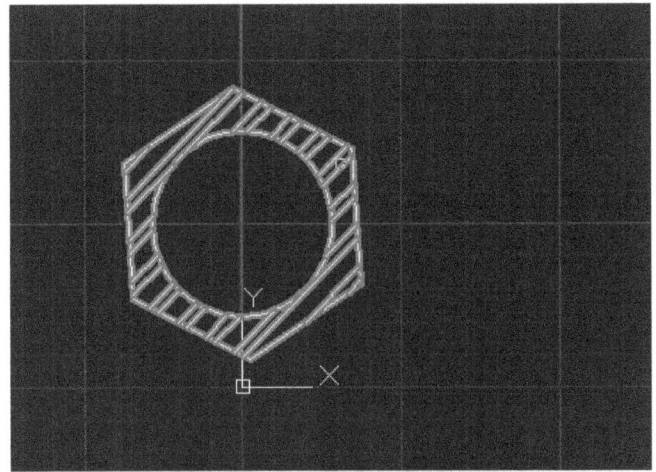

Pic 3.201 Click on object

4. Specify the base point.

```
Specify base point or [Displacement] <Displacement>: click
```

5. For example, I use center of my circle as base point.

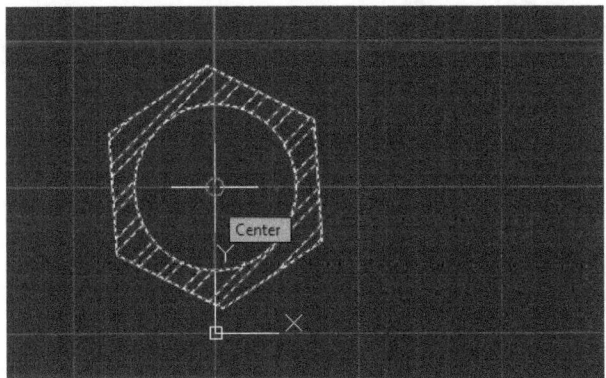

Pic 3.202 Click on object's center

6. Specify second point where you want the the base coordinate to be moved into.

```
Specify second point or <use the first point as displacement>:
```

7. When you want to click, you can see the preview of the object.

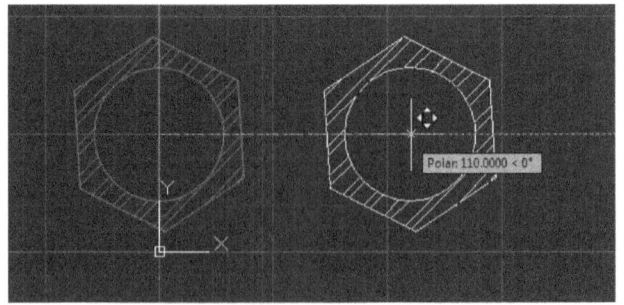

Pic 3.203 Initial and final position

8. Click Enter, the object will have new position.

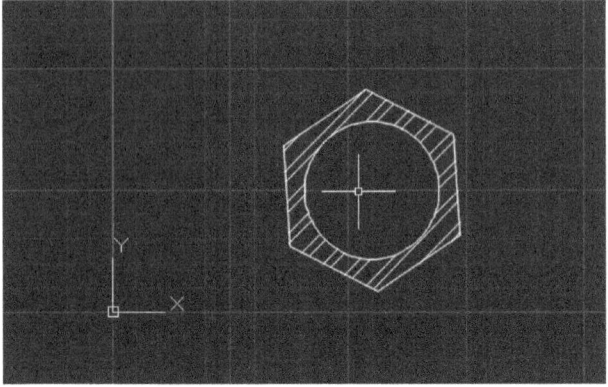

Pic 3.204 New object's position

3.3.2 Rotate

Rotate command rotates object based on base point and degrees of rotation. See example below:

1. For example, I have object on following pic:

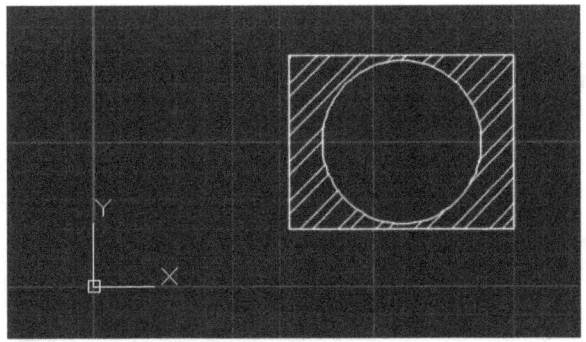

Pic 3.205 Object to rotate

2. Type rotate, then select object you want to rotate.

```
Command: ROTATE
Current positive angle in UCS:  ANGDIR=counterclockwise   ANGBASE=0
Select objects: 3 found, 1 group
Select objects: [click on object]
```

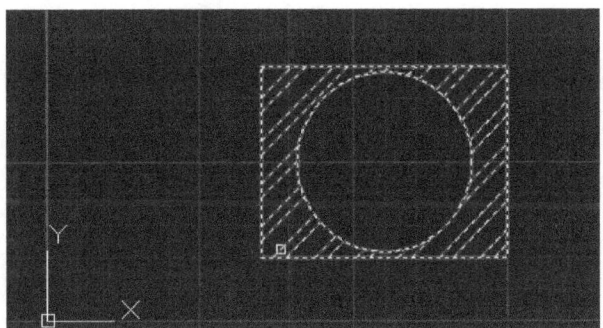

Pic 3.206 Selecting the object

3. Specify the base point for rotation.

```
Specify base point: [click on base point]
```

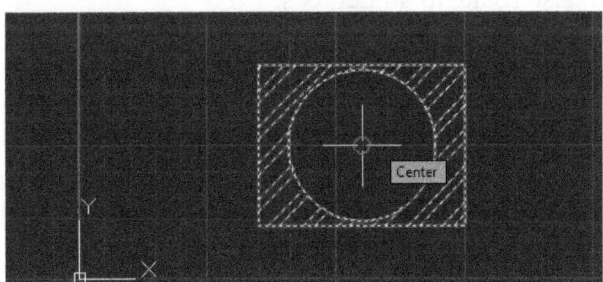

Pic 3.207 Specify the base point

4. If already clicked, rotation icon appears.

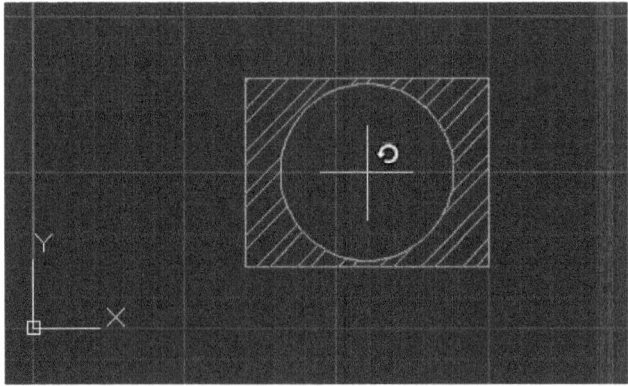

Pic 3.208 Rotation icon appears

5. Set the rotation degrees to -45.

```
Specify rotation angle or [Copy/Reference] <0>: -45
```

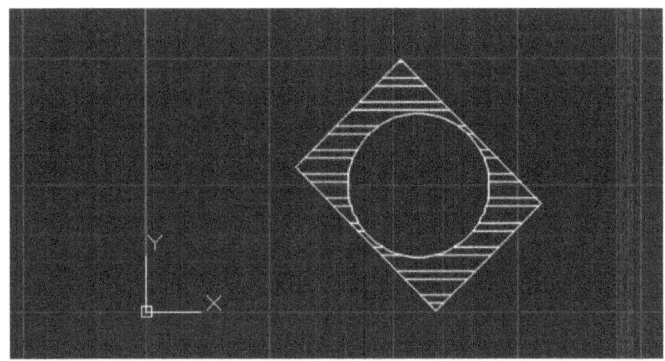

Pic 3.209 Object after rotated to -45 using the center of circle as base point

6. You can also rotate to 90 degrees:

```
Command: ROTATE
Current positive angle in UCS: ANGDIR=counterclockwise  ANGBASE=0
Window Lasso  Press Spacebar to cycle options3 found, 1 group
Select objects:
Specify base point:
Specify rotation angle or [Copy/Reference] <45>: 90
```

7. The object will be rotated by 90 degrees.

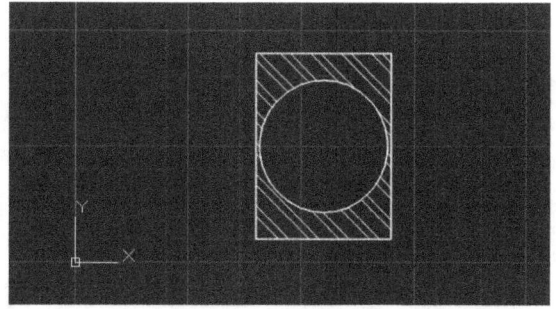

Pic 3.210 Object 90 degrees rotated

3.3.3 Trim

Trim command trims certain part of objects. See steps below to see the example of TRIM function:

1. For example there is three circle objects.

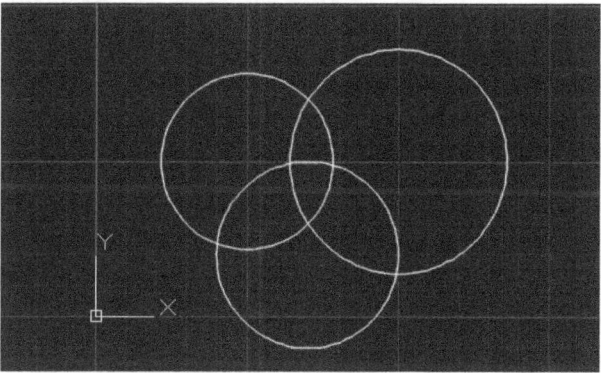

Pic 3.211 Three circle objects

2. You'll trim the inside part of the intersection. Type "trim first".

3. Select all objects.

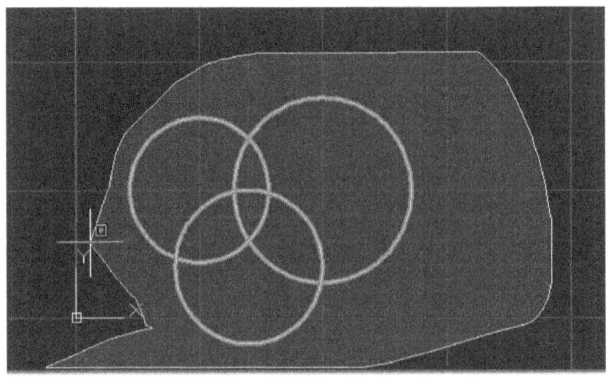

Pic 3.212 Select all objects

4. Selected objects will becoming dotted line.

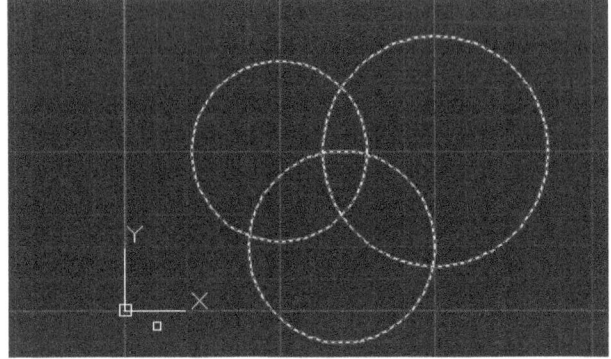

Pic 3.213 Selected objects in dotted line

5. Click on the segments you want to trim.

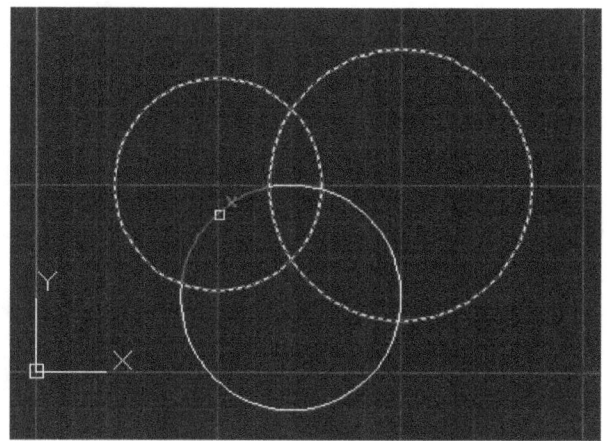

Pic 3.214 Click on segments you want to trim

6. The segment you click will disappear/trimmed. If ERASE erase all of the object, trim will erase selected segment of object.

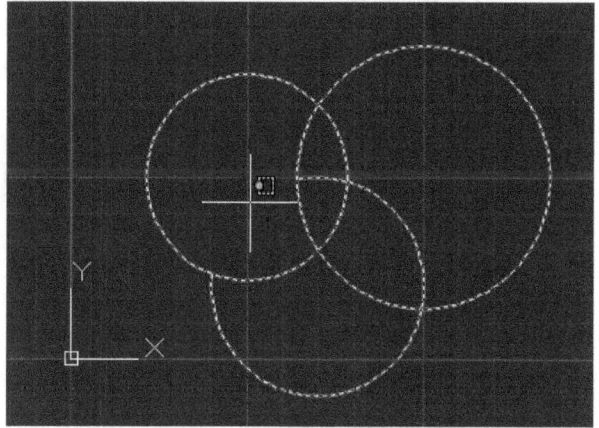

Pic 3.215 Trimmed segment disappears

7. You can click other segment to trim.

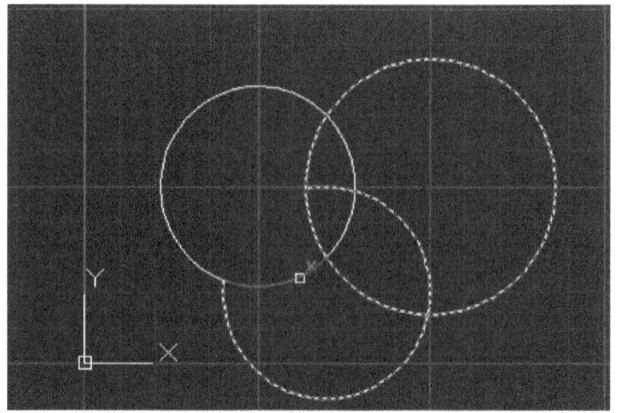

Pic 3.216 Clicking other segment to trim

8. The other segment will diasppear too.

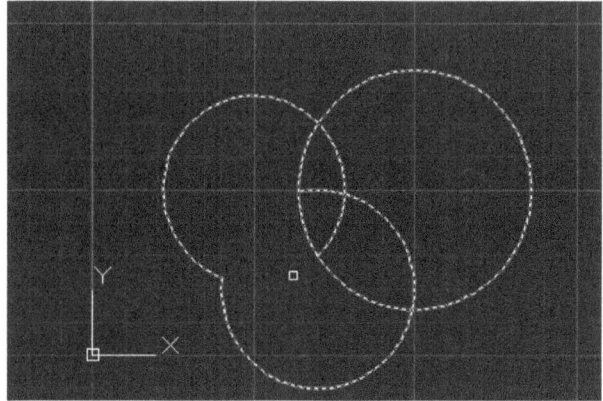

Pic 3.217 Second segment disappears

9. You can click other segment to trim it.

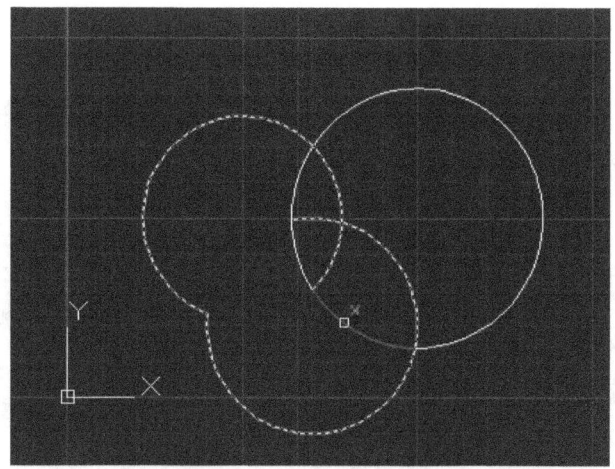

Pic 3.218 Selecting the segment

10.   Final result will be as below:

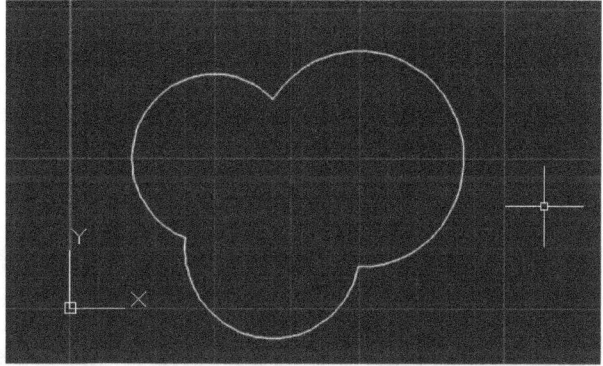

Pic 3.219 Result of trim process

✓ *Exercise Drawing with The Trim Command*

The Trim command allows you to remove additional lines up to an intersection point. You can also switch to Erase within the Trim command by typing R. This removes lines that do not intersect. This removes lines that do not intersect, such as the Delete command. Start the Trim command and press Enter to select the entire sketch for trimming. Trim the overhanging lines as shown in the image. If you have accidentally removed a line, type "U" to undo it. Also take a close look at any lines stuck between the small edges. These will most likely cause problems in the extrusion process that turns your sketch into 3D. Press Enter to confirm when you are done.

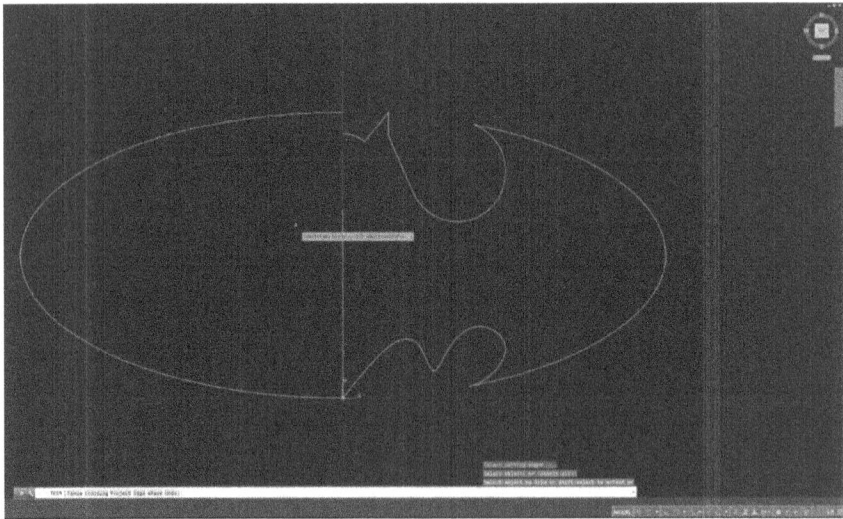

Next, mark the line in the middle and the free ellipse on the left and delete it. Finally, select the small ellipse line in the upper triangle and delete it as well.

After trimming and erasing you should get this.

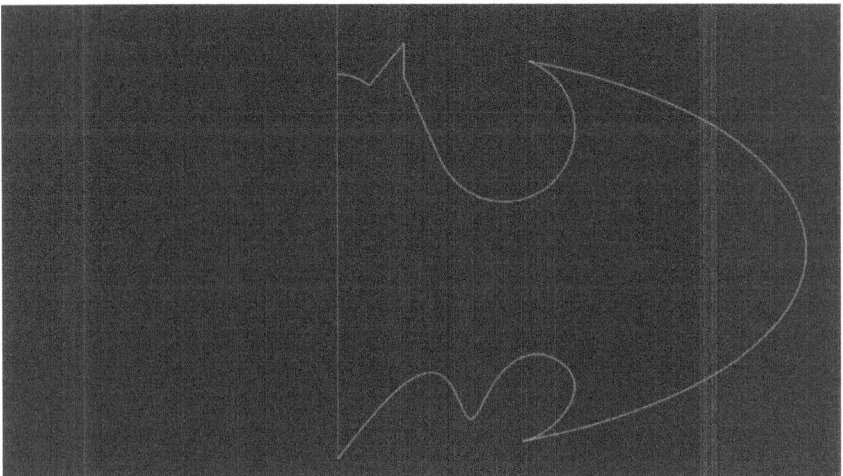

3.3.4 Extend

Extend command extends line or the arc to certain object. See example below for more advanced:

1. For example, there are one the arc and one line. The arc is going to be extended to the line.

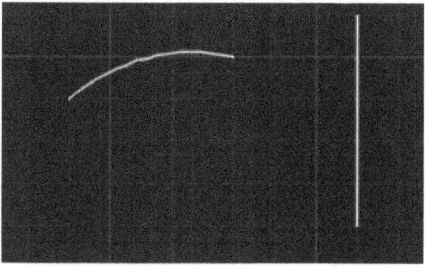

Pic 3.220 An arc and a line

2. Type "extend" command.
3. Select all objects.

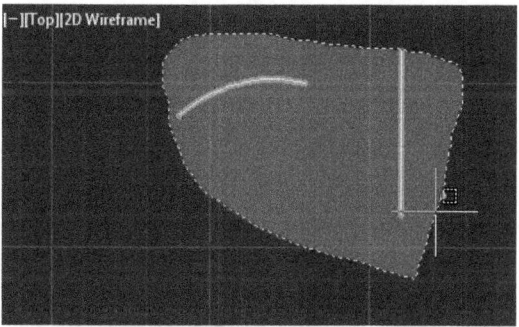

Pic 3.221 Selecting all objects

4. Both objects becoming dotted line.

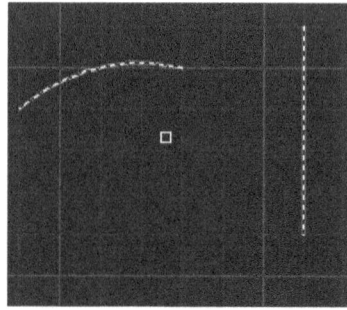

Pic 3.222 Selected objects becoming dotted line

5. Click on object you want to extend, the object become extended.

```
Select object to extend or shift-select to trim or
[Fence/Crossing/Project/Edge/Undo]:
```

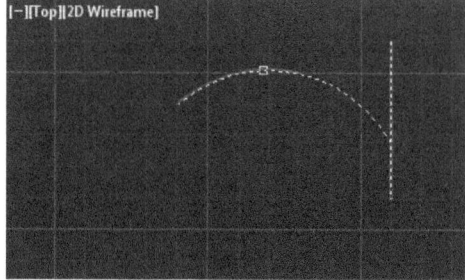

Pic 3.223 Object extended

6. See following pic for the result.

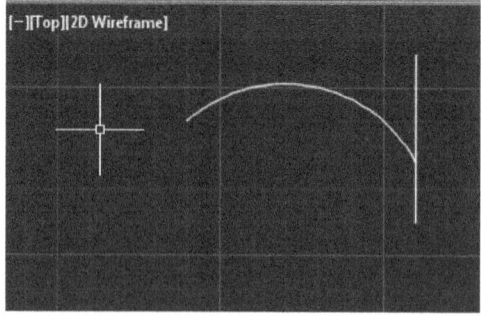

Pic 3.224 The arc has been extended

3.3.5 Erase

Erase command erases selected object. Erase will erase all part of selected object, not only the segments. Here's how to use erase object:

1. From picture below, the hatch will be erased.

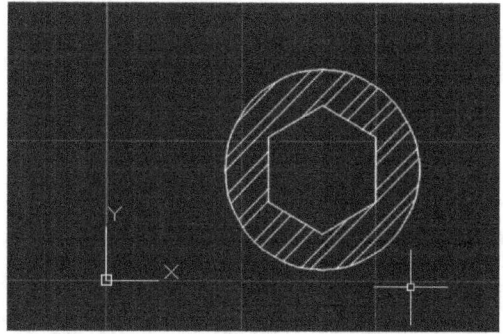

Pic 3.225 The hatch erased

2. Type "erase" in document.

3. Pointer icon will be changed to erase mode.

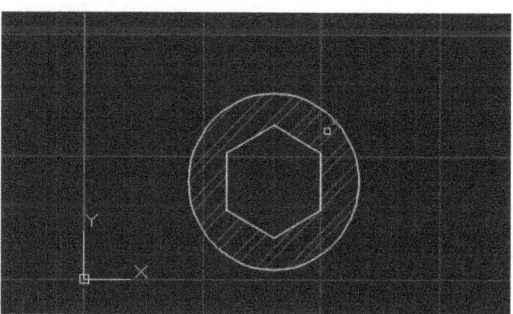

Pic 3.226 Pointer ready to erase

4. Click on the hatch to erase and click **Enter**.

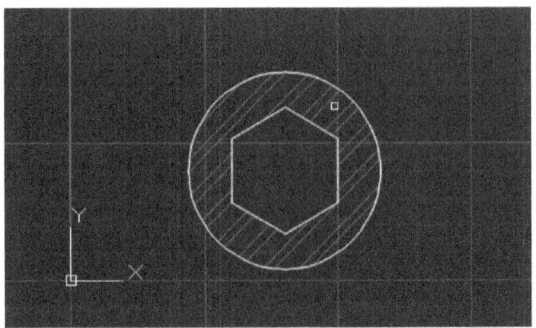

Pic 3.227 Click on the hatch

5. The hatch will be erased.

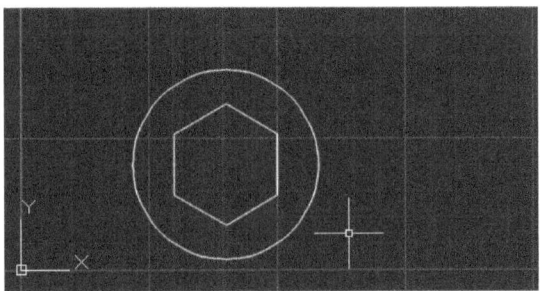

Pic 3.228 Object after the hatch erased

3.3.6 Copy

Copy command copies object, where the copied object still exist. See example below:

1. Type "Copy".

2. Choose the object to copy.

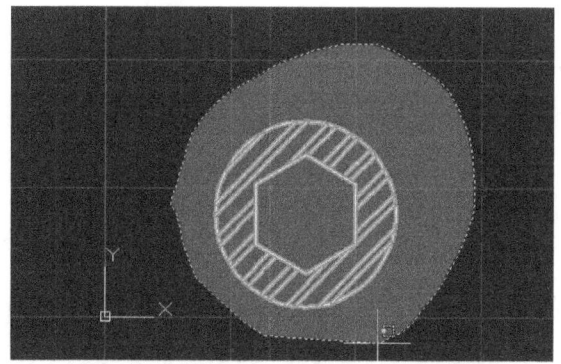

Pic 3.229 Selecting object to copy

3. Select base point.

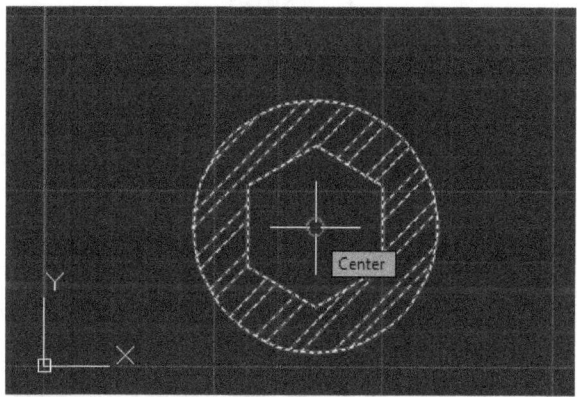

Pic 3.230 Selecting base point

4. Show the new position, you can use polar or relative coordinate.

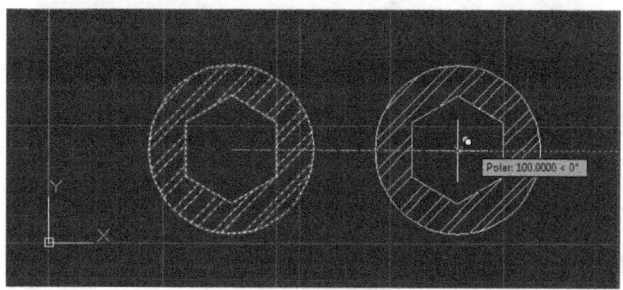

Pic 3.231 Specifying the new position

5. The copying result will be displayed in AutoCAD. And the initial object still exists.

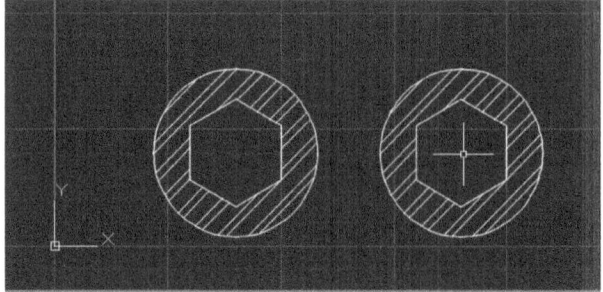

Pic 3.232 Copying result

3.3.7 Mirror

Mirror command mirrors object using a line as the mirror. See steps below for mirror command example:

1. Type "mirror' command.
2. Select object you want to mirror.

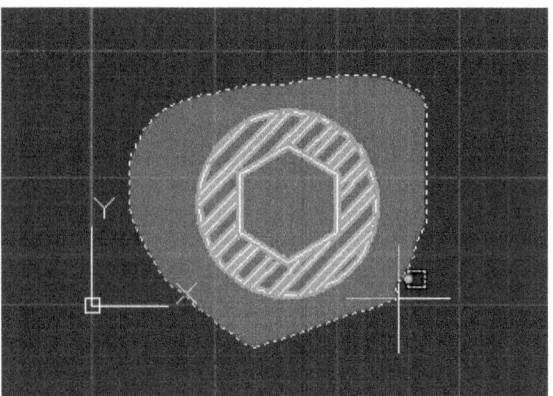

Pic 3.233 Selecting object

3. Selected object will become dotted line.
4. Specify the first line to make the mirror.

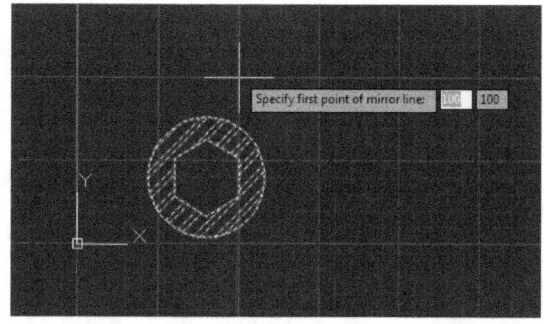

Pic 3.234 First line for the mirror

5. Specify the second line for the mirror.

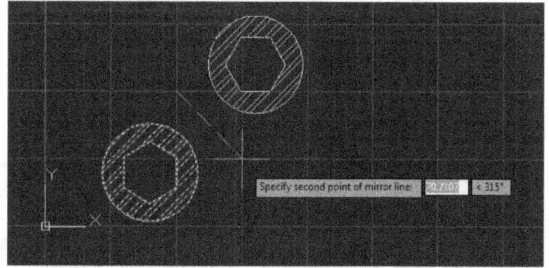

Pic 3.235 Specify the second line for the mirror

6. The object will be mirrored, and you'll be asked whether you want to initial object or not?

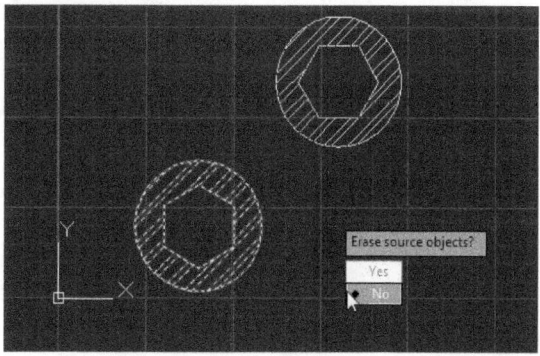

Pic 3.236 Option to erase initial object or not

7. You can see the initial object and mirrored object on the drawing area.

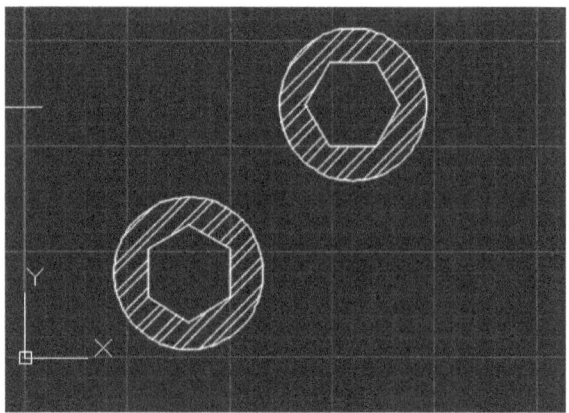

Pic 3.237 Mirroring area

3.3.8 Fillet

The fillet can be made from two line, see example below:

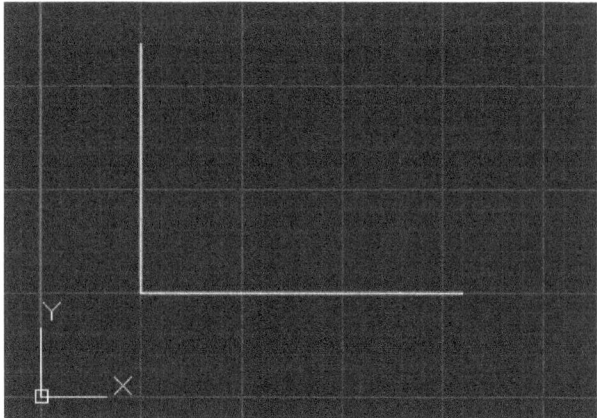

Pic 3.238 Line to be filleted

See example below on how to create fillet:

1. Run "fillet" command.

```
Command: FILLET
Current settings: Mode = TRIM, Radius = 0.0000
```

2. Click R and set fillet radius to 40.

```
Select first object or [Undo/Polyline/Radius/Trim/Multiple]: R
Specify fillet radius <40.0000>: 40
```

3. Click on the first line.

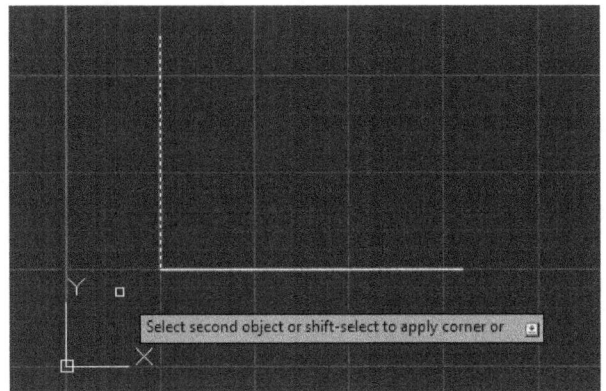

Pic 3.239 Click on the first line

4. Click on the second line.

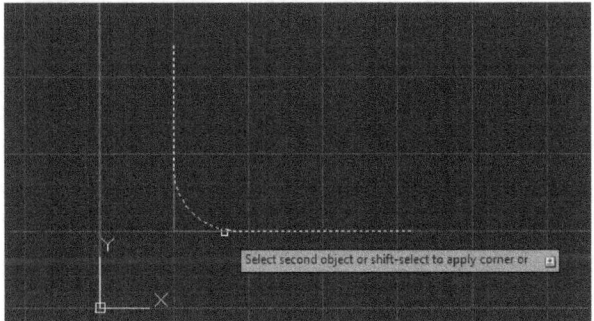

Pic 3.240 Second fillet

5. After you click the second line, fillet created automatically.

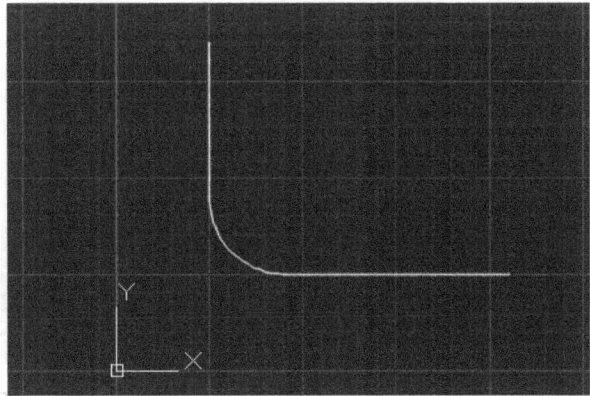

Pic 3.241 Fillet result

3.3.9 Chamfer

Chamfer similar to fillet, but chamfer is not an the arc, it's a line. See example below to create chamfer:

1. Execute "chamfer" command, click D to specify the distance of the chamfer.

```
Command: _chamfer
(TRIM mode) Current chamfer Dist1 = 0.0000, Dist2 = 0.0000
Select first line or
[Undo/Polyline/Distance/Angle/Trim/mEthod/Multiple]: D
```

2. Set first distance to 40, and second distance to 40.

```
Specify first chamfer distance <0.0000>: 40
Specify second chamfer distance <40.0000>: 40
```

3. Click on the first line.

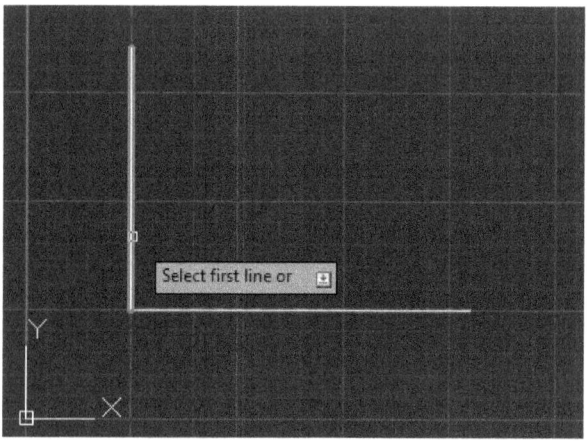

Pic 3.242 Click on the first line

4. Click on the second line.

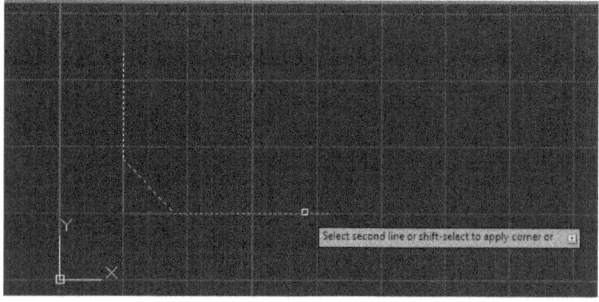

Pic 3.243 Click on the second line

5. See picture below for the chamfer result.

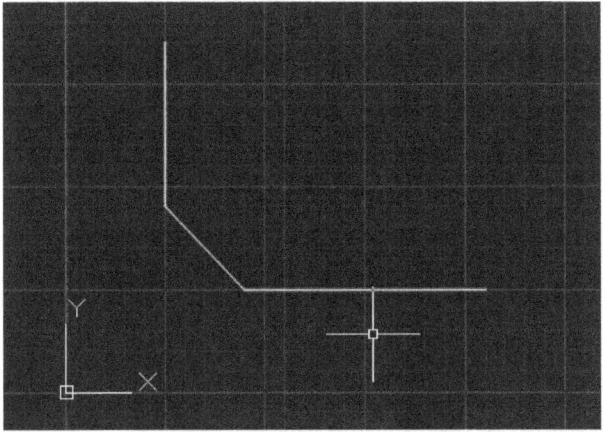

Pic 3.244 Chamfer result

3.3.10 Explode

Explode command explodes polyline or region to segments. See steps below:

1. There's a polyline:

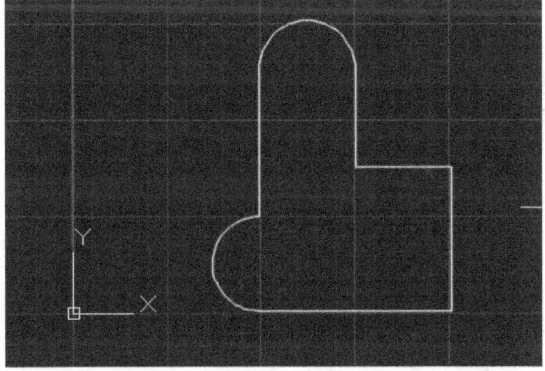

Pic 3.245 Polyline

2. If you choose a polyline, all segments will become a dotted line, this is because it's one object.

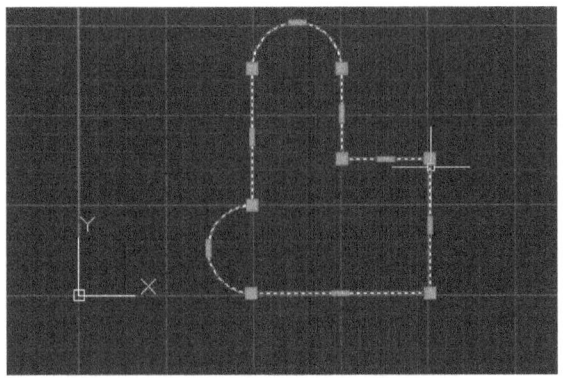

Pic 3.246 All segments of polyline become dotted line

3. Now execute "explode" function.

4. Select the polyline object.

Pic 3.247 Polyline selection

5. Click Enter, the object will be exploded. If you click on the object, a segment will be selected. This means the object already segmented/exploded.

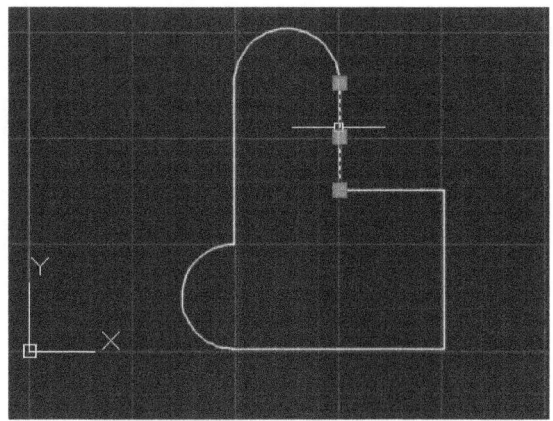

Pic 3.248 The object after segmentation

6. If you want to choose more than one segment, you have to click those segments, one by one.

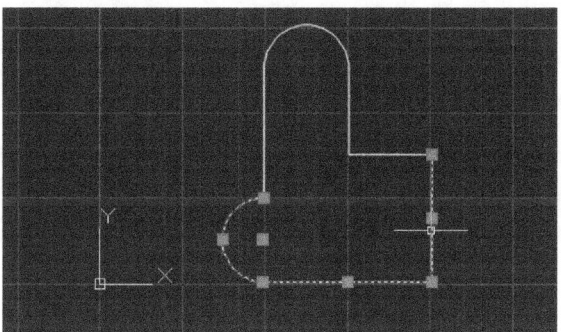

Pic 3.249 Choosing 3 segments after exploded

3.3.11 Stretch

Stretch command stretches object. You just have to define which object to stretch, see example below:

1. For example, there is an object as below:

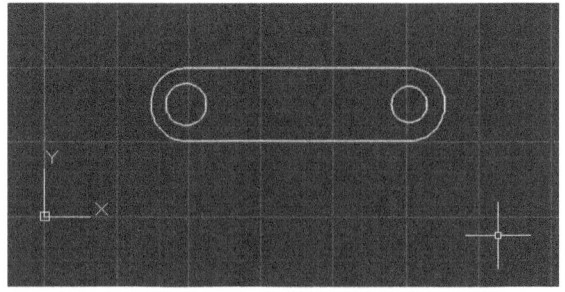

Pic 3.250 Object to stretched

2. Execute "stretch" command, and select part of the object you want to stretch.

```
Command: STRETCH
Select objects to stretch by crossing-window or crossing-polygon...
```

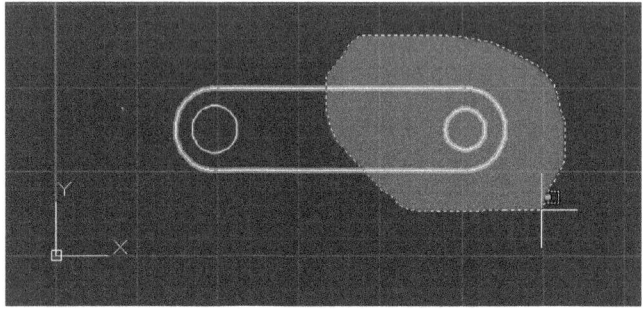

Pic 3.251 Choosing object to stretch

3. Selected object will be a dotted line.

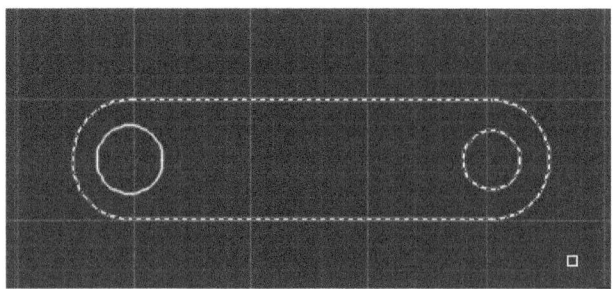

Pic 3.252 Selected object becoming dotted line

4. Click Enter, and specify the base point.

```
Specify base point
```

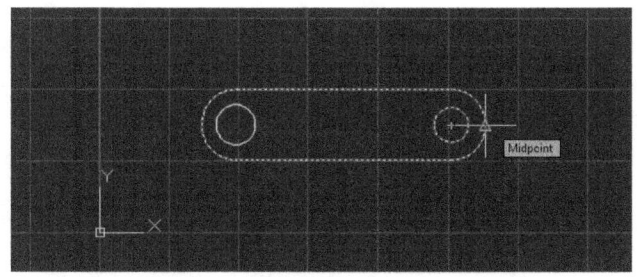

Pic 3.253 Specify base point for stretching

5. Click on a base point and click the second point.

```
Specify base point or [Displacement] <Displacement>:
Specify second point or <use the first point as displacement>:
```

6. Drag to right, you can see the initial position and position after stretching.

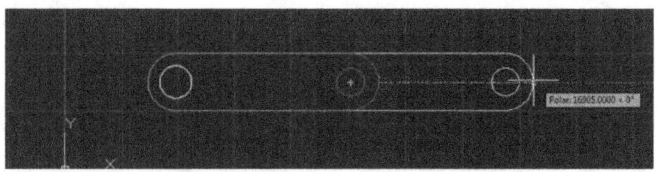

Pic 3.254 Stretching to right

7. If the mouse drag released, the object will be stretched.

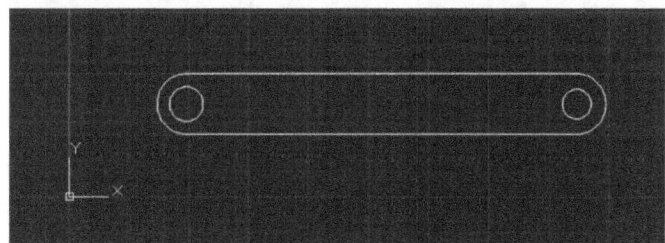

Pic 3.255 The object after stretching

8. Stretch can also be used for makes size smaller. By dragging to the left.

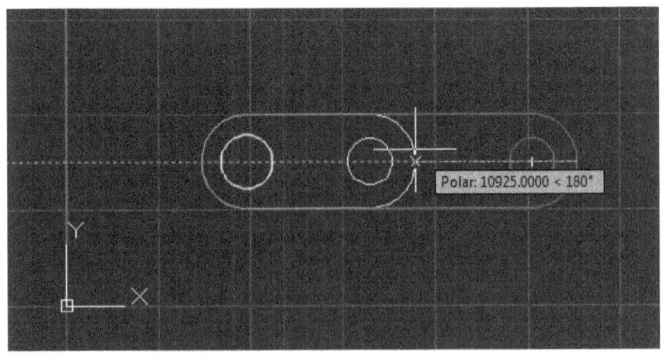

Pic 3.256 Negative stretching

9. If you do negative stretching, the object will be smaller.

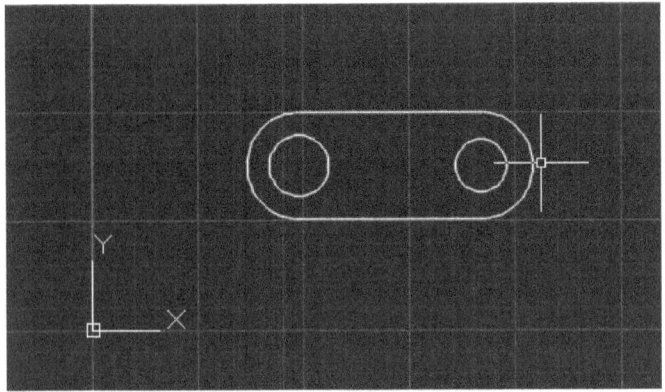

Pic 3.257 Negative stretch makes the object smaller

3.3.12 Scale

Scale command will scales object to make the object larger or smaller. See example below:

1. Type "scale".

2. Select the object.

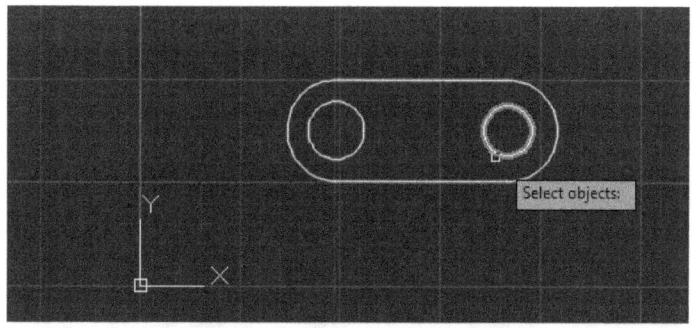

Pic 3.258 Choose the object to scale

3. Click Enter, the object will be a dotted line.

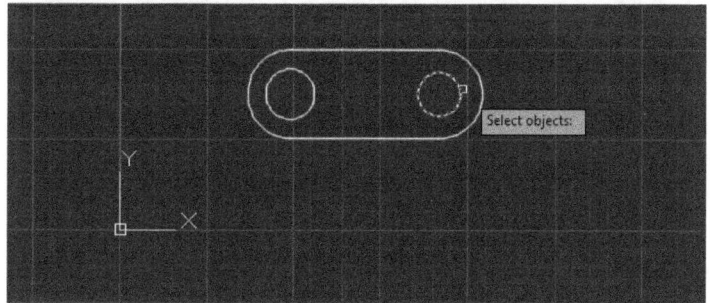

Pic 3.259 Object selected

4. Specify base point for scaling.

```
Specify base point
```

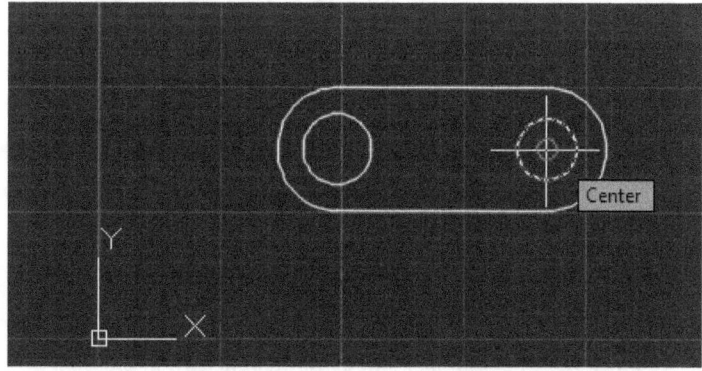

Pic 3.260 Click on the center as base point for scaling

---

5. Specify the scale factor or zoom factor, for example, if I take 2, it means the object will be zoomed twice.

```
Specify scale factor or [Copy/Reference]: 2
```

6. The result is:

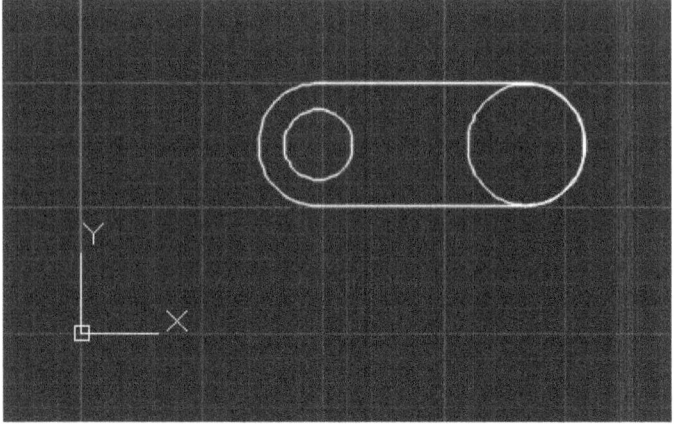

Pic 3.261 The result of scale factor

3.3.13 Array Rect

You can copy object and paste it in array of rows and columns by using array rect. This is how to use Array Rect command:

1. Type "arrayrect" in the command prompt.

2. Select the object.

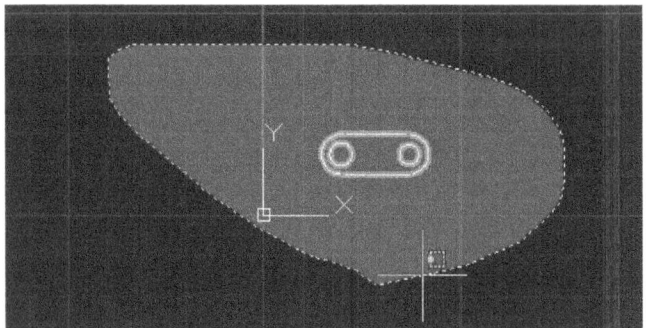

Pic 3.262 Select object you want to copy with array rect

3. Object automatically copied with array rect.

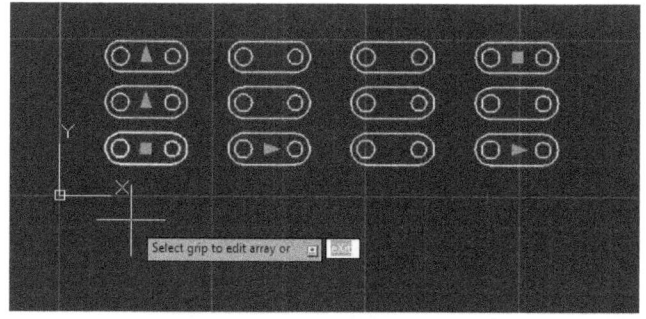

Pic 3.263 Object copied as array

4. You can change the property of array rect using column and row in **Columns** and **Rows**.

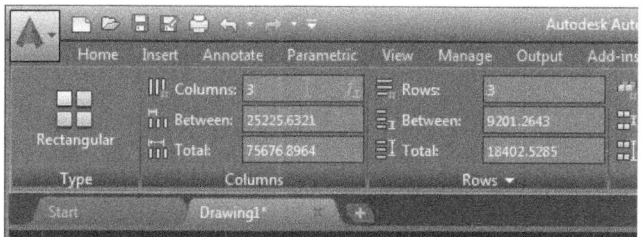

Pic 3.264 Columns and rows

5. Click **Close Array** to close the array creation.

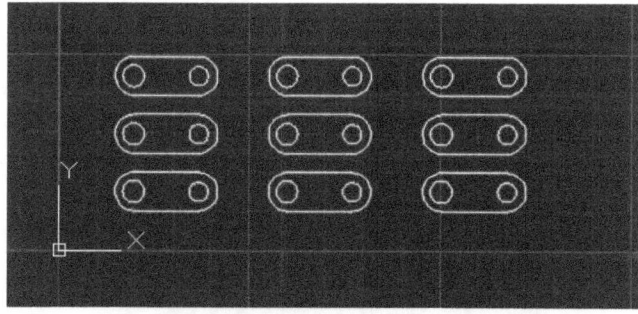

Pic 3.265 Array rect result

CHAPTER 4 CASE STUDIES

On this chapter, I'll demonstrate how to implement skills you have learned from the previous chapter to draw a simple drawing.

4.1   Create Simple House Plan

For example, you will create simple house plan with size 100x100. See steps below:

1. Set limits from workspace from 0,0 to 100, 100.

```
Command: LIMITS
Reset Model space limits:
Specify lower left corner or [ON/OFF] <0.0000,0.0000>: 0,0
Specify upper right corner <100.0000,100.0000>: 100,100
```

2. Draw a line as below:

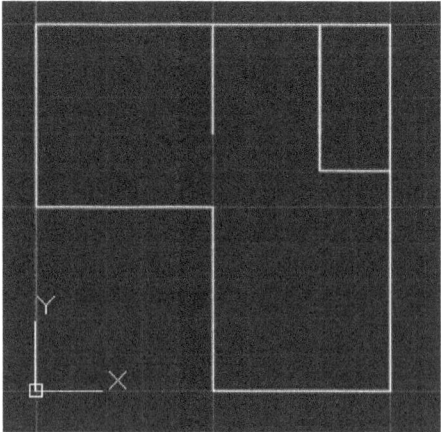

Pic 4.1 Drawing a line

3. See pic above, the size is 100 x 100.
4. Create small rectangle with size = 2.5 x 2.5.

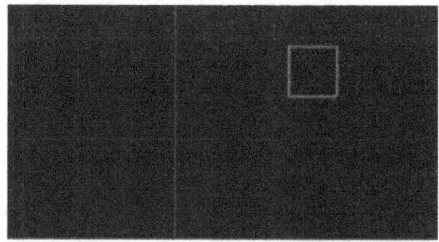

Pic 4.2 Small rectangle

5. Type Move, and click the object.

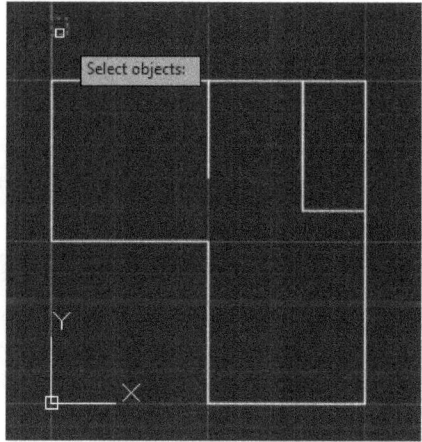

Pic 4.3 Small object

6. Choose the midpoint of the little rectangle as a base point.

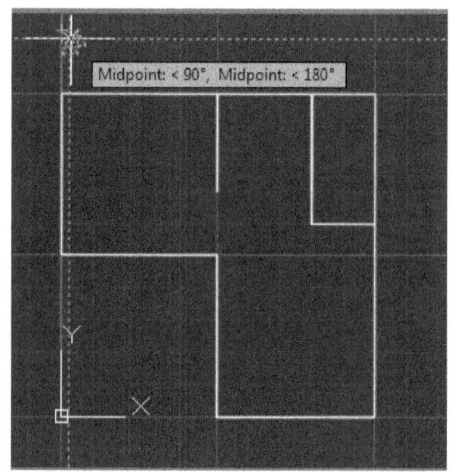

Pic 4.4 Choosing base point of move

7. Put the small rectangle to the corner of each line.

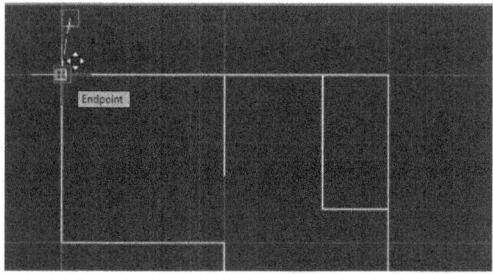

Pic 4.5 Put small rectangle to the corner

8. The box will be available in the corner.

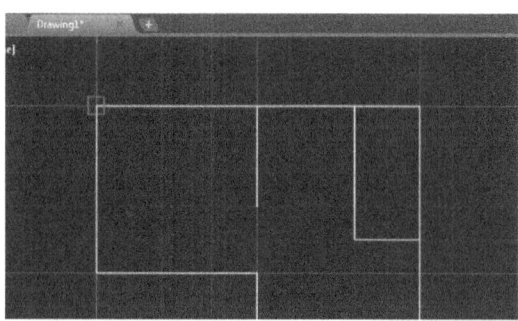

Pic 4.6 Box in the corner

9. Type "Copy" and select the small object.

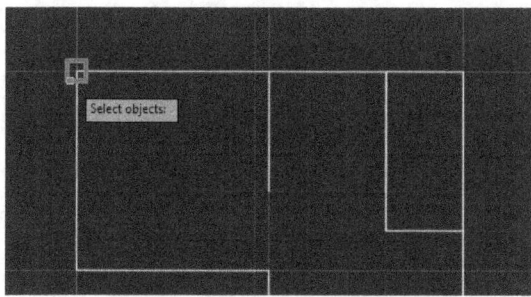

Pic 4.7 Choose the small rectangle to copy

10. The small rectangle will become dotted line.

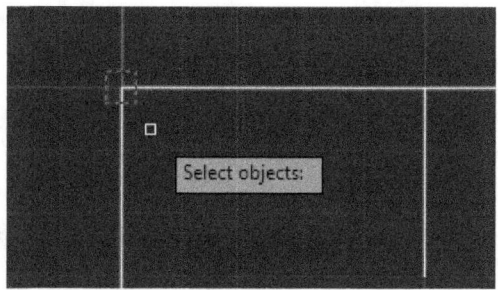

Pic 4.8 Small rectangle selected

11. Click on the mid rectangle when you are asked: **Specify base point**.

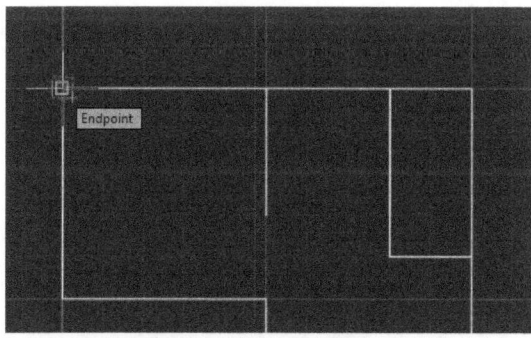

Pic 4.9 Specify base point for copying

12. Then choose other corner/intersection point in **Specify end point**,

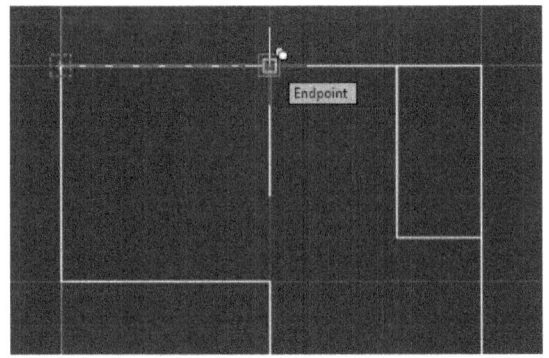

Pic 4.10 Specify the end point

13. Do this in each intersection/corner.

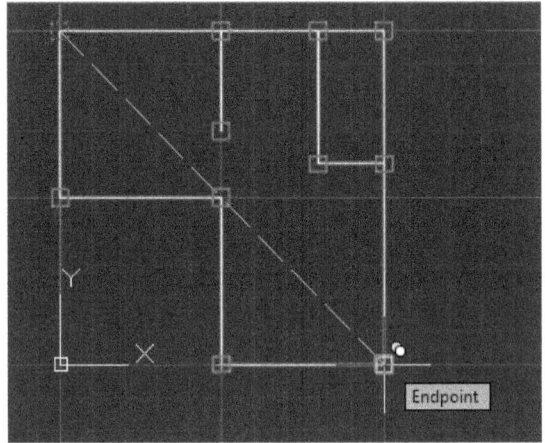

Pic 4.11 Copying the small rectangle in all corner/intersection

14. The result will be like the picture below:

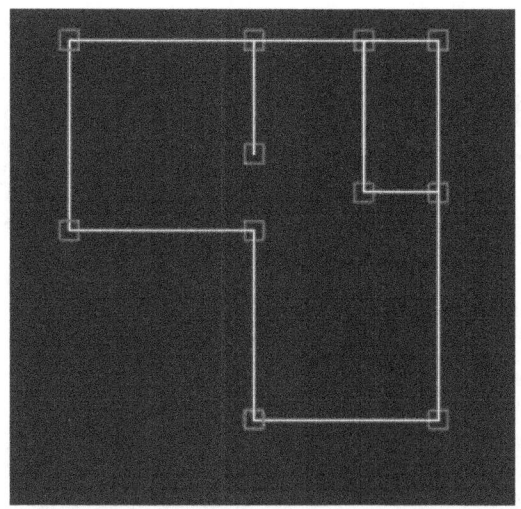

Pic 4.12 Small rectangle copied to each corner

15. Draw a line to draw the wall.

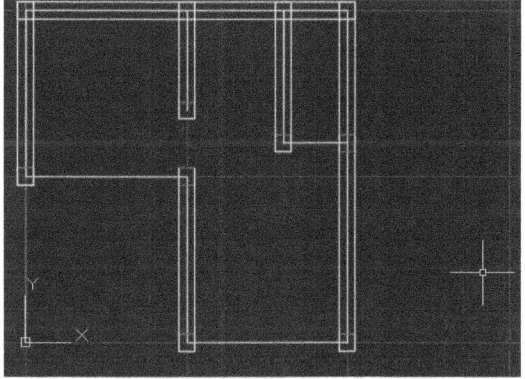

Pic 4.13 Drawing the wall

16. You can also draw a line to make a border.

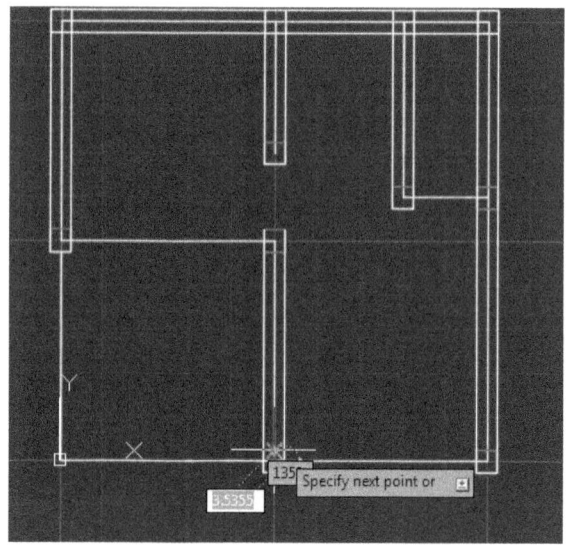

Pic 4.14 Drawing a line to draw border and wall

17. Draw a door like this.

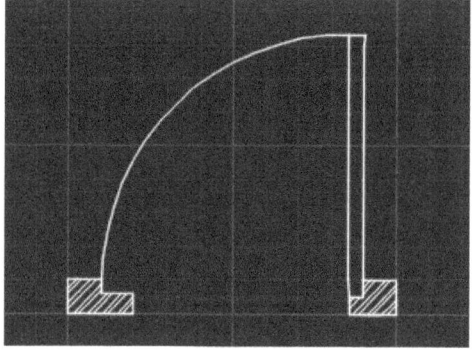

Pic 4.15 Drawing a door

18. Move the door to the place you want to create door.

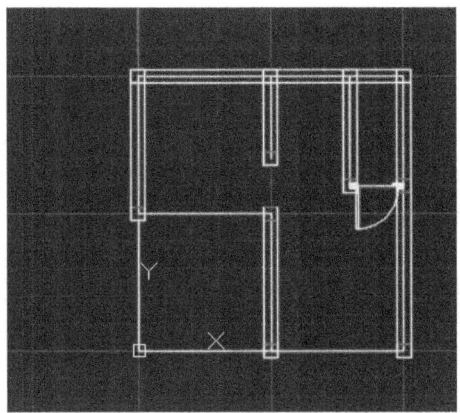

Pic 4.16 Create door

19. To give grass effect, create hatch, and select the pattern to Grass.

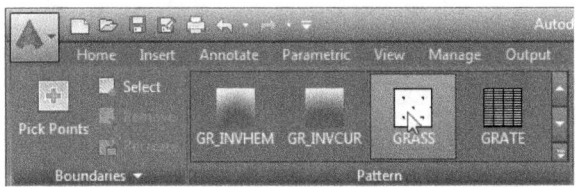

Pic 4.17 Choose pattern to grass

20. Give grass to the area you want to draw a grass.

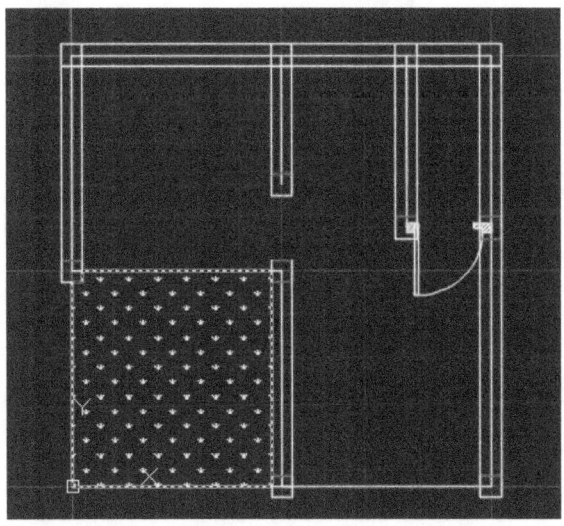

Pic 4.18 Giving grass by hatching using pattern = grass

21.   To give a car, click on **View > Tool Palettes**.

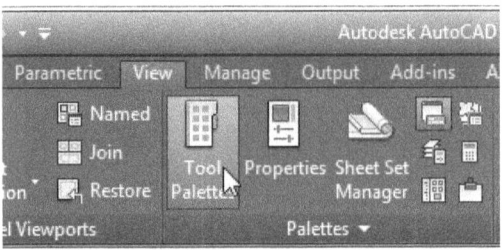

Pic 4.19 Click on View > Tool Palettes

22.   Choose **The architectural > Vehicles**. Right click and select **Properties**.

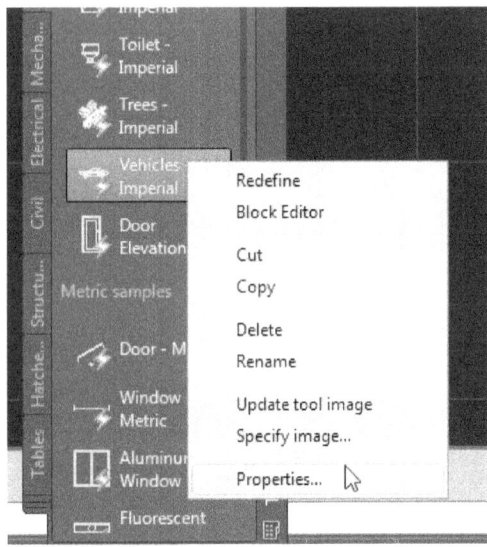

Pic 4.20 Click Properties

23.   Choose Type (view) to Sports Car (Top).

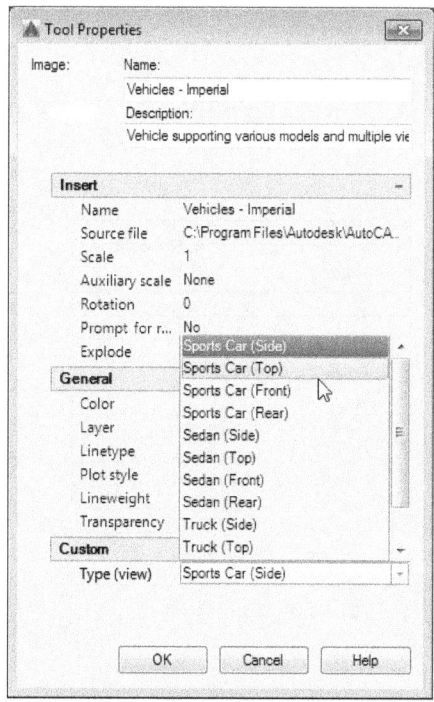

Pic 4.21 Choosing Sports Car (Top)

24. You can see the Type (View) changed and click OK

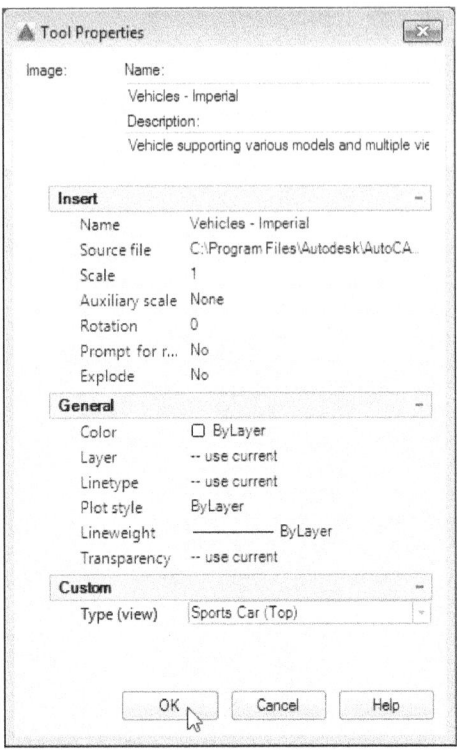

Pic 4.22 Type (view) property for car object already changed

25. Click to insert car object.

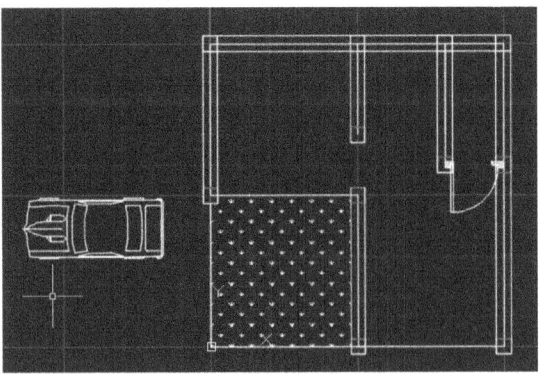

Pic 4.23 Car object inserted to drawing

26. Rotate using rotate function and put it in the garage.

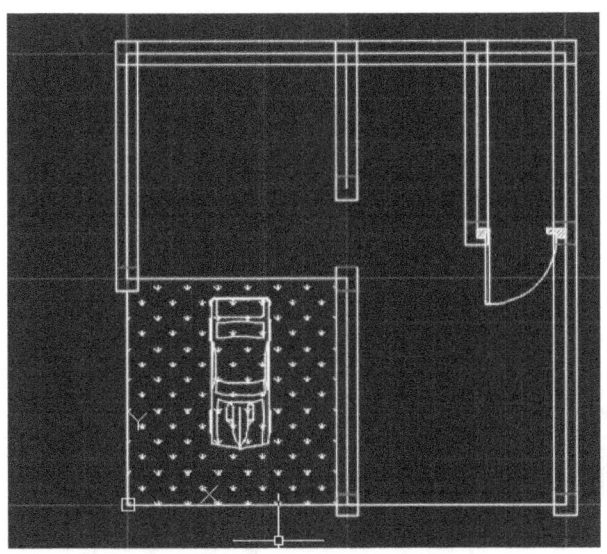

Pic 4.24 Put the car object

27. Using the same method, you can add other objects, like tree.

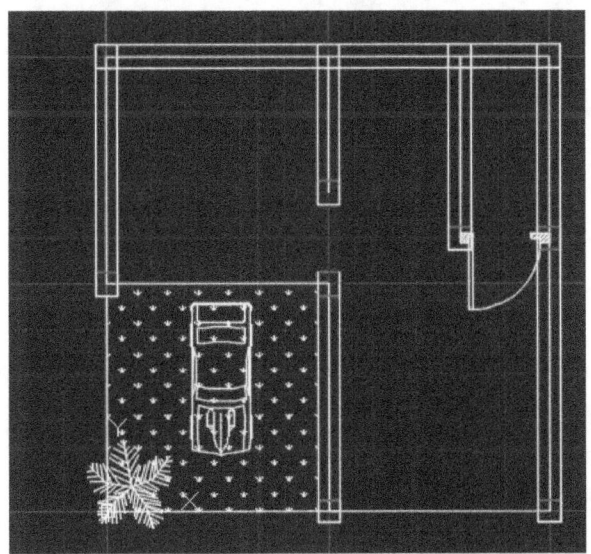

Pic 4.25 Inserting another object

28. To insert annotations, click **Home > Annotation**.

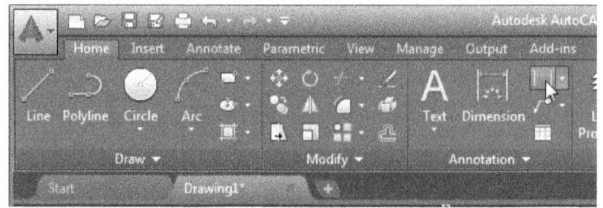

Pic 4.26 Annotation box

29. Complete the annotation in another place.

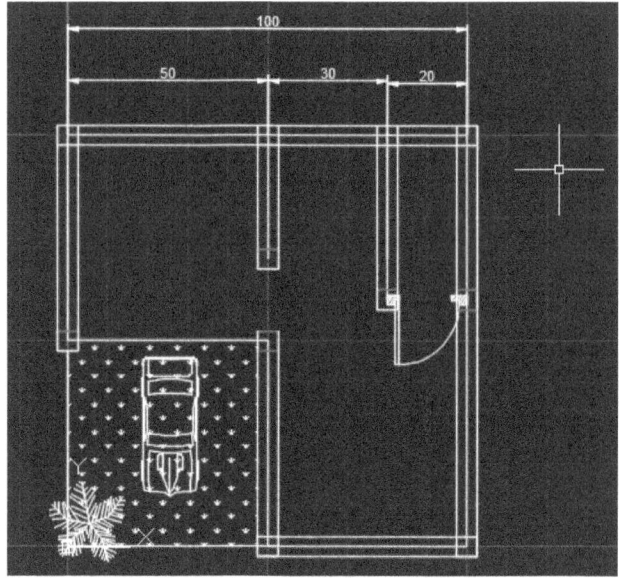

Pic 4.27 Completing the annotation

30. You can create other objects to complete the drawing using polyline, circle and rectangle.

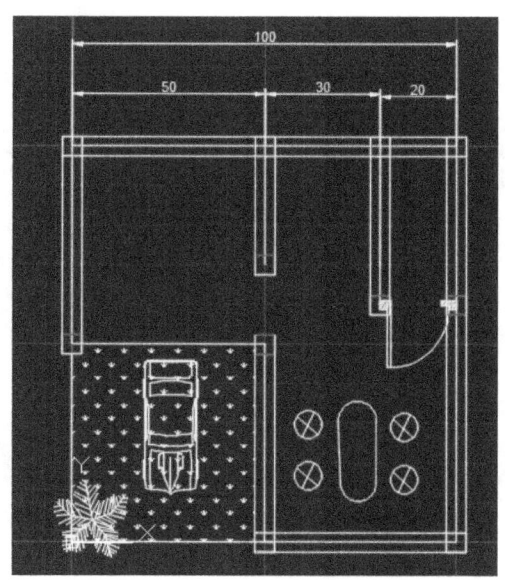

Pic 4.28 Completing the object

31.   You can again add more annotation.

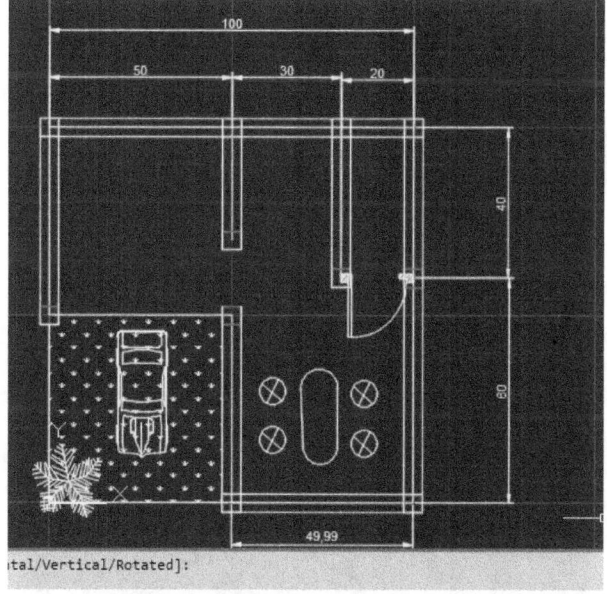

Pic 4.29 Adding annotations on other places

32.   The result will be like this, you can add more using your
creativity.

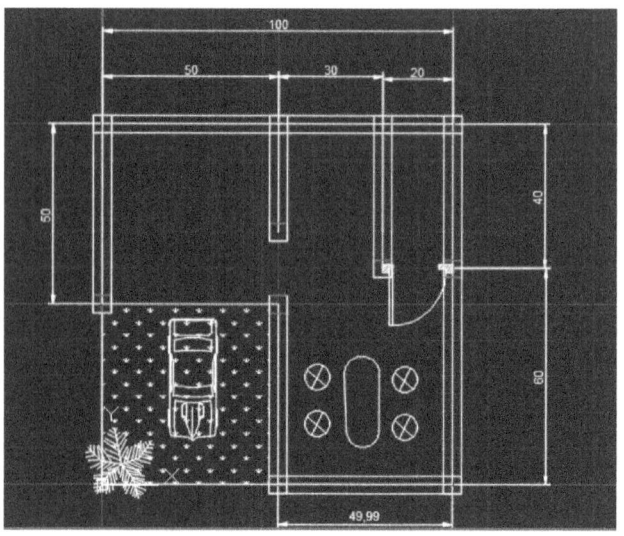

Pic 4.30 Final result of house plan drawing

4.2 Create Simple Gear

In this tutorial, you'll learn about how to create simple gear, follow steps below:

1. Create two circles, with an identical center point, but with different radius. Then add the teeth of the gear.

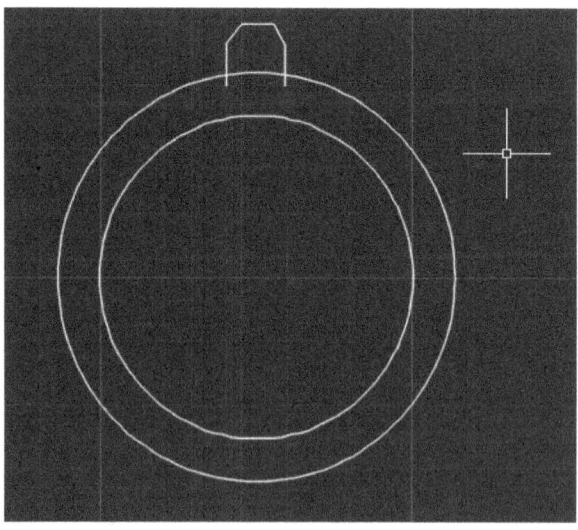

2. Trim the root of the teeth by entering Trim command then select all objects.

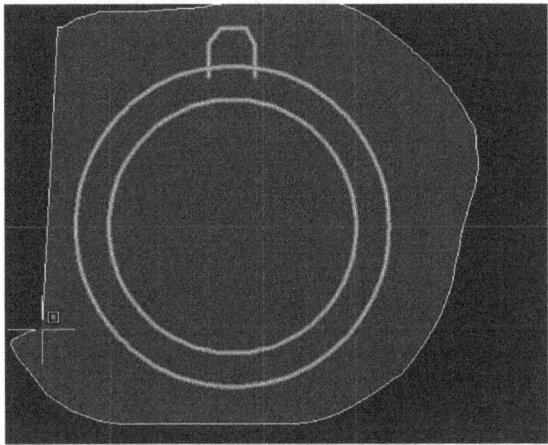

Pic 4.32 Selecting all objects to trim

3. Click on the root of the teeth to trim it.

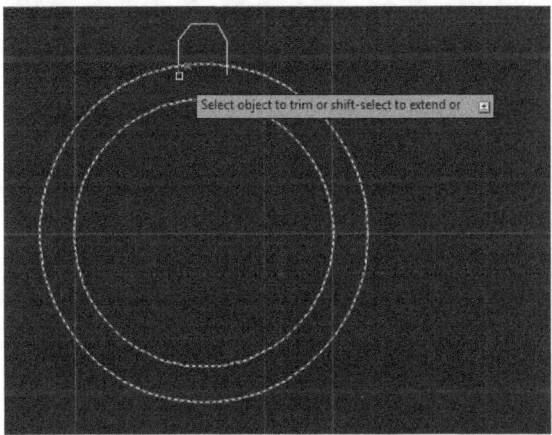

Select object to trim or shift-select to extend or

Pic 4.33 Trim the tooth's root

4. You can see the tooth.

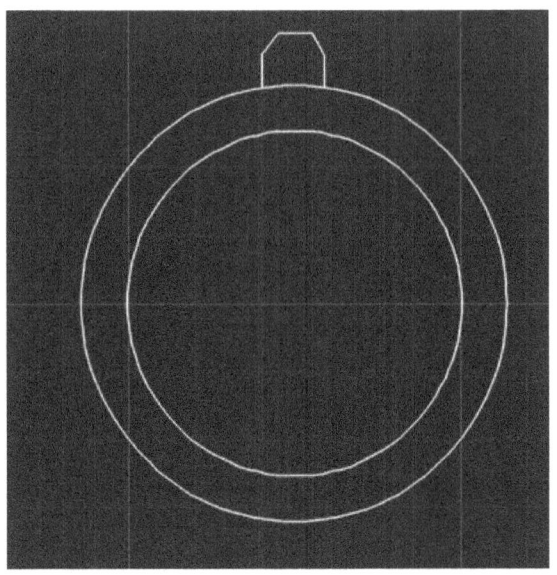

Pic 4.34 Circle with one tooth

5. Copy the object 18X, execute copy command, and select objects.

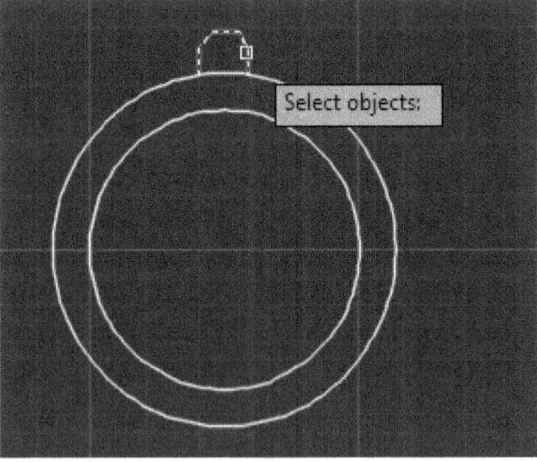

Pic 4.35 Selecting object to copy

6. Then rotate the copied tooth using Rotate command, choose the object then specify a base point to the center.

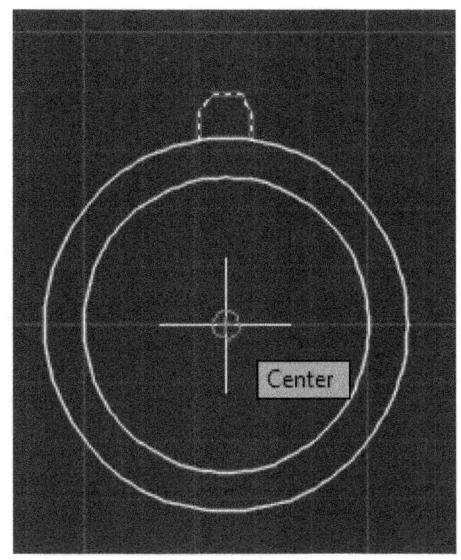

Pic 4.36 Specify base point = the center

7. Rotate with 20 degrees interval.

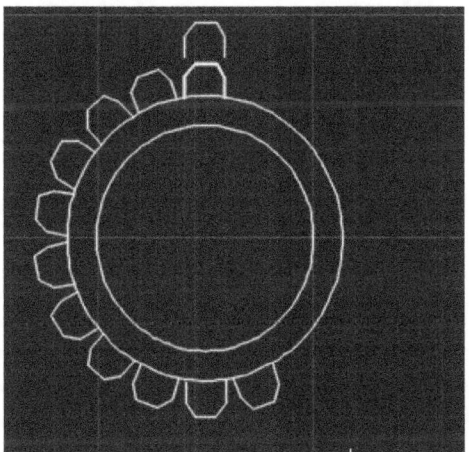

Pic 4.37 Rotating the teeth with 20 degrees' interval

8. Do until all the teeth rounding the circle.

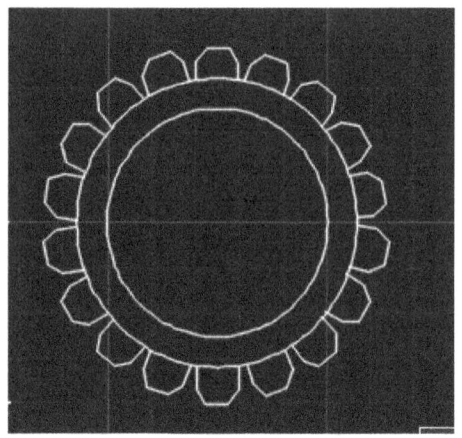

Pic 4.38 Teeth rounding the circle

9. Type "linetype", and click Load.

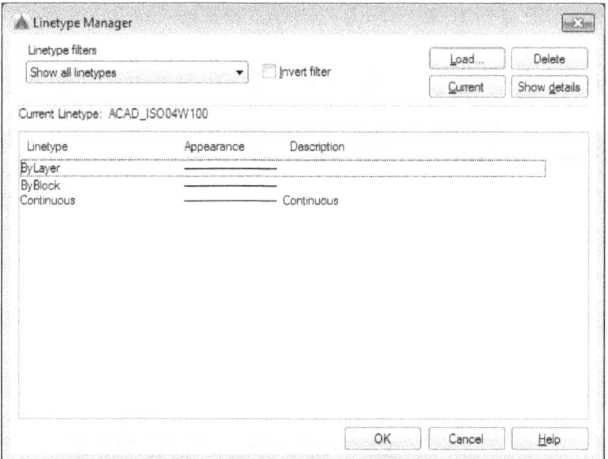

Pic 4.39 Choosing the linetype

10. Choose ISO long-dash dot to draw the axis.

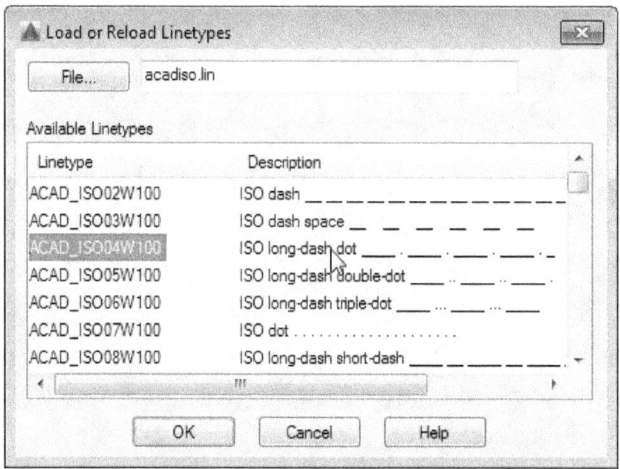

Pic 4.40 Adding ISO long-dash dot

11.   Then click on the iso long-dash dot, and click Load.

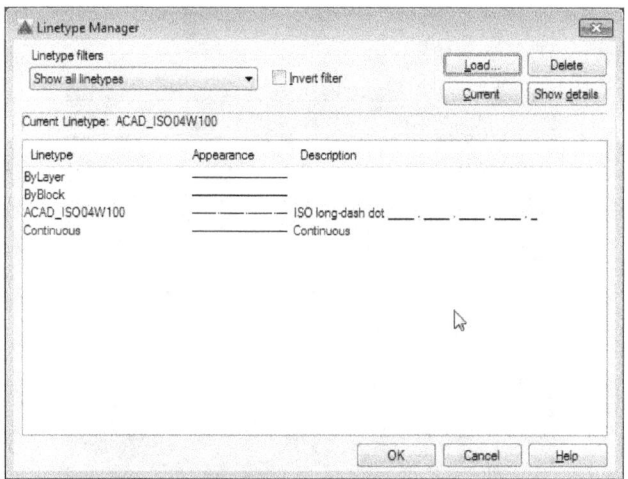

Pic 4.41 Choosing Longtype dash dot

12.   Draw vertical axis.

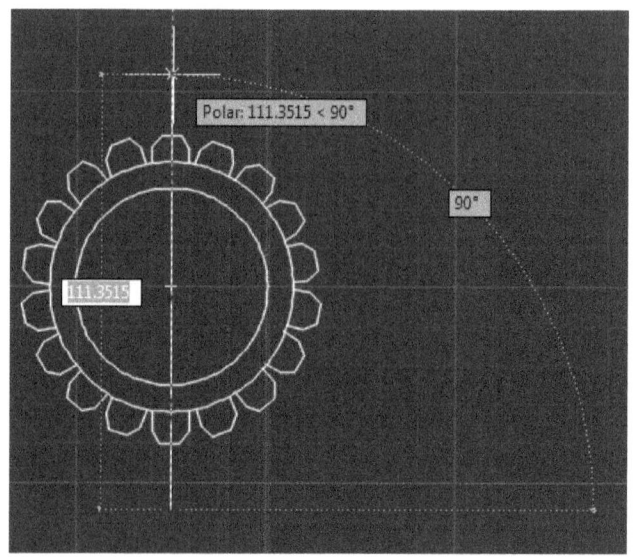

Pic 4.42 Drawing vertical axis

13.   Drawing horizontal axis.

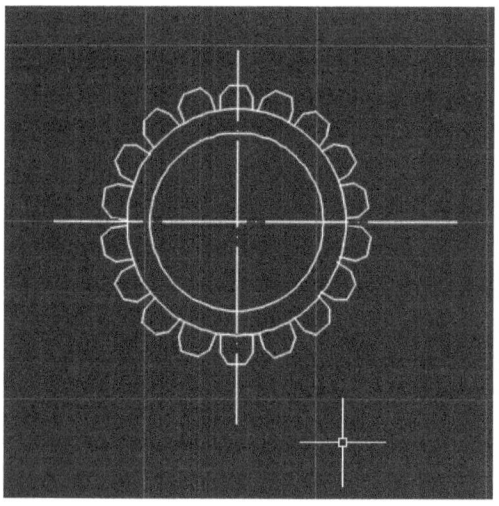

Pic 4.43 Drawing horizontal axis

14.   Trim the gear, and select all the objects.

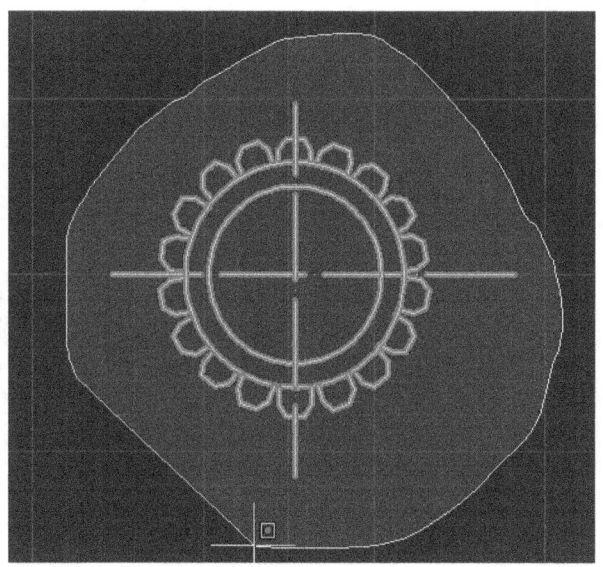

Pic 4.44 Select the gear

15.   Click on the line below the teeth.

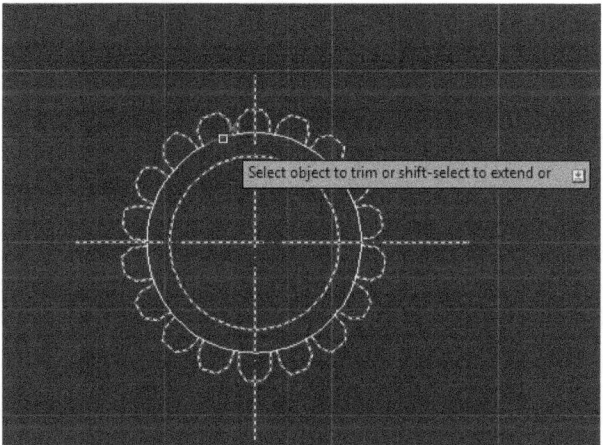

Select object to trim or shift-select to extend or

Pic 4.45 Trimming the line below the teeth

16.   The final result will be as below:

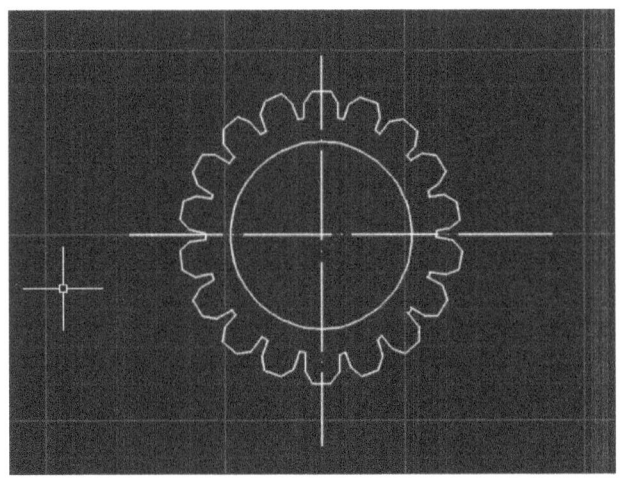

Pic 4.46 Final result creating gear

4.3 Create Simple Piston

See example below for creating simple piston using AutoCAD:

1. Create two circles, and two lines.

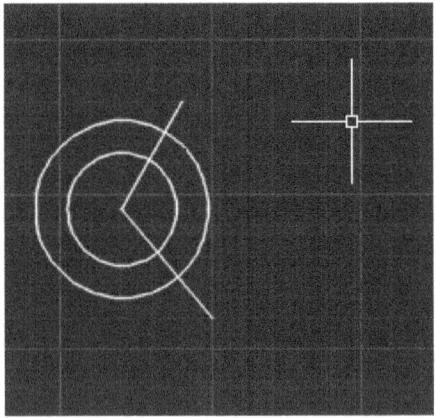

Pic 4.47 Create two circles and two lines

2. Type "Trim" and select all objects.

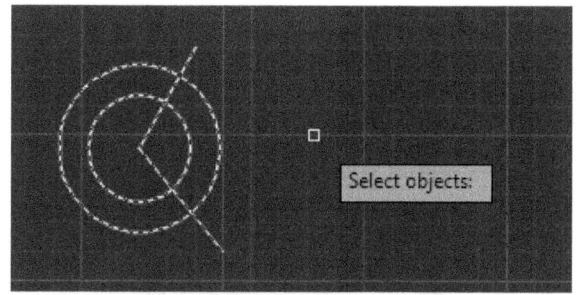

Pic 4.48 Choosing all objects to trim

3. Trim to make picture below:

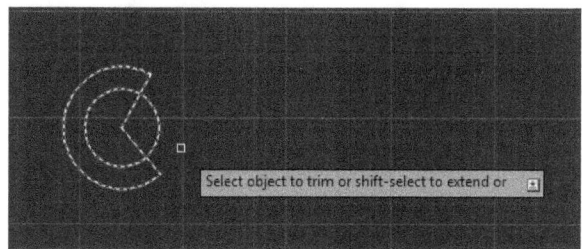

Pic 4.49 Trim outer circle

4. Trim part of the inner circle, see picture below:

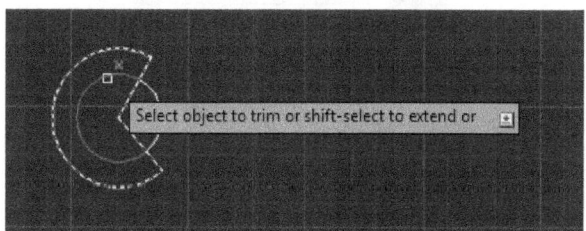

Pic 4.50 Trim inner circle

5. Trim the radius line.

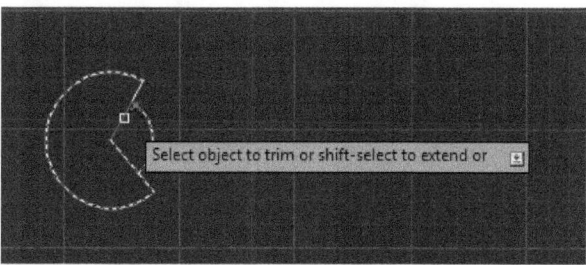

6. The result will be like this.

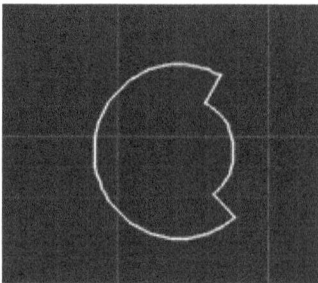

Pic 4.52 Engine axle drawing

7. Draw a small circle, with center point identical with the center point of the axle.

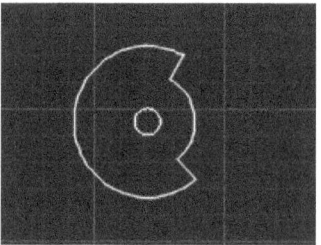

Pic 4.53 Small circle

8. Create one more small circle.

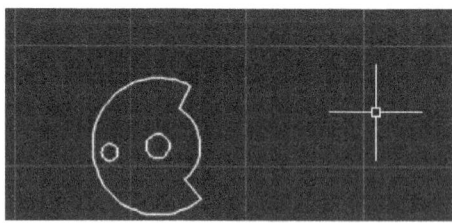

Pic 4.54 Creating one more small circle

9. Create polyline like picture below:

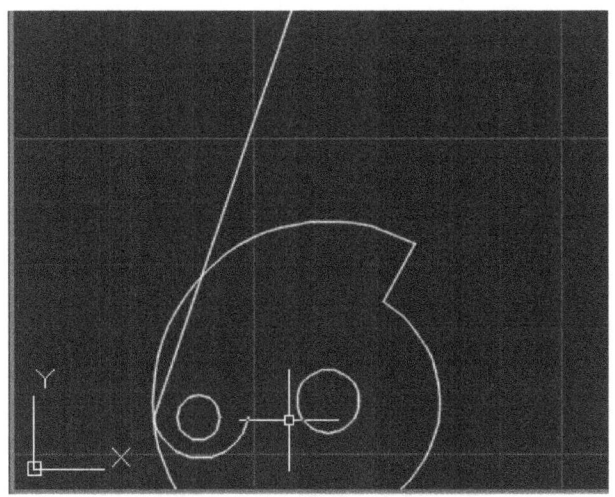

Pic 4.55 Create polyline

10.   Add the polyline with line and the arc.

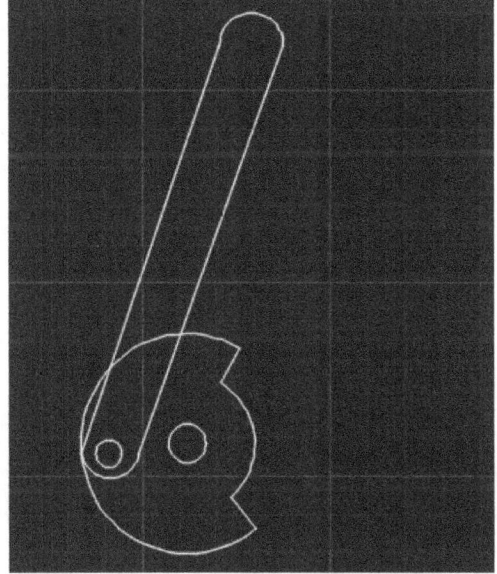

Pic 4.56 Create polyline with line and the arc

11.Create polyline to create the piston.

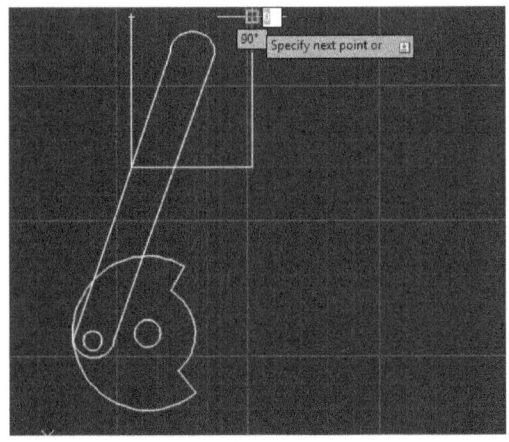

Pic 4.57 Create polyline to draw the piston

12.   Draw an the arc to form the top of the piston.

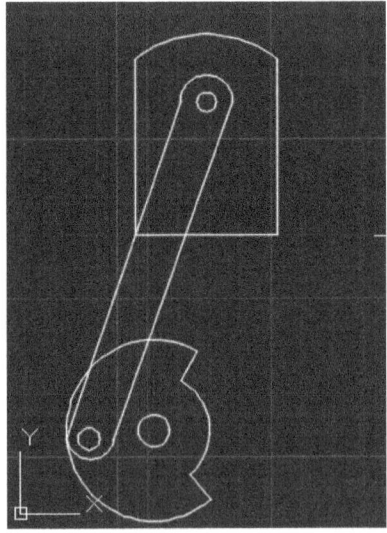

Pic 4.58 Draw an arc to form the top of piston

13.   Now use trim.

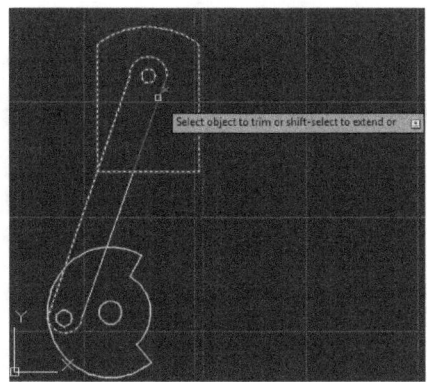

Pic 4.59 Trimming

14.   The result after trim look as following picture.

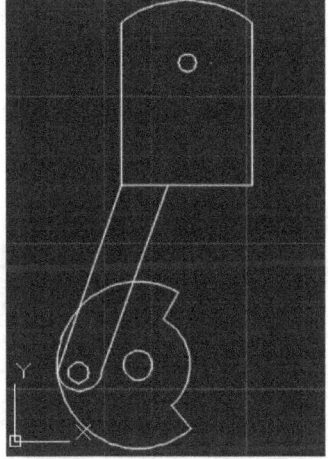

Pic 4.60 Result after trimming

15.   Create two rectangles to draw the piston rings.

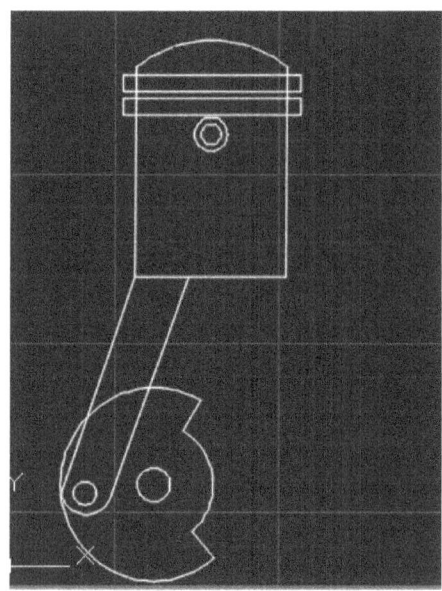

Pic 4.61 Create two rectangles

16.   Type trim, and select to trim.

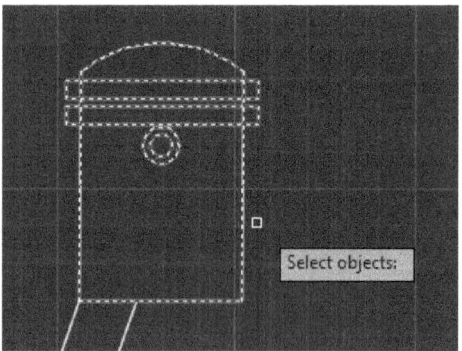

Pic 4.62 Select objects to trim

17.   Trim on the wall side of the piston inside the piston ring.

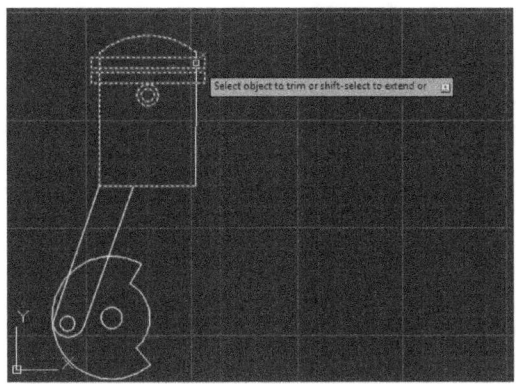

Pic 4.63 Trimming on piston ring

18. The final result is:

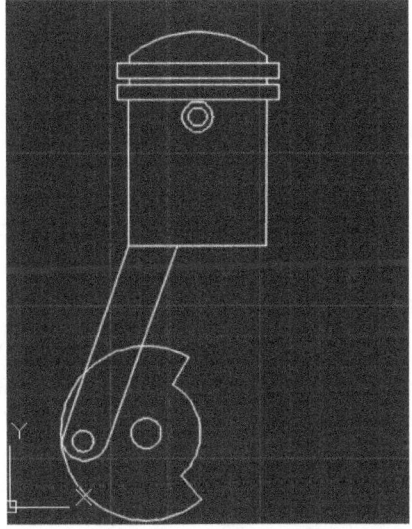

Pic 4.64 Final result

CHAPTER 5 DRAW 3D DRAWING

This is probably the most interesting part of this AutoCAD tutorial for beginners – we're nearing 3D designing! In this chapter, you will learn how to create basic 3D in a 3D modeling workspace. You can use three-dimensional (3D) shapes of solid objects to create boxes, cones, cylinders, spheres, rings, discs, and pyramids.

To create a Solid 3D, change the custom 3D modeling workspace to create and modify a Solid 3D model. At the end of this chapter, we also explained some keyboard shortcuts to work more effectively with AutoCAD. When working in 3D, you should remember, that drawing in AutoCAD is only possible on the XY-Plane. If you want to change the direction to draw or plot your 3D object, you must redefine the coordinate system. Draw a random circle in your DrawSpace while being in Top view. Now enter Front view and type "UCS". This will allow you to set a new coordinate system. Type in "V" to set your current view as the new coordinate system. Draw a second circle concentric with the first one. Now rotate the model by holding Shift and the mouse wheel, and you will see the 3D alignment of both circles.

5.1   Configure 3D Workspace

Do steps below for configuring 3D workspace:

1. In status bar, click on Workpsace Switching.

Pic 5.1 Workspace Switching

2. In the menu, click 3D Basics

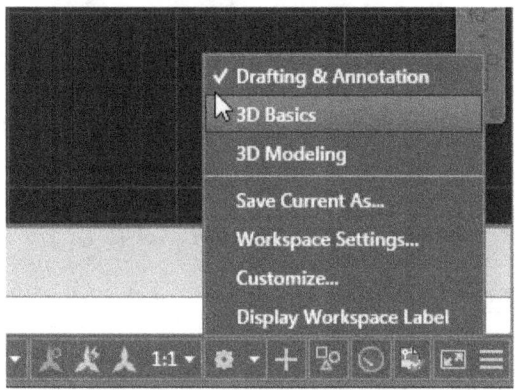

Pic 5.2 Clicking 3D Basics

3. 3D Basics workspace displayed, you can access lots of command and tools for drawing 3D objects.

5.2   Draw 3D Objects

Similar to 2D drawing, there are some basic objects in 3D Drawing. You'll learn how to draw 3D objects below:

5.3.2 Draw Box

Box is a rectangle with height. Here are steps to create box in atuocad:

1. Click Box icon on icon Create toolbar.

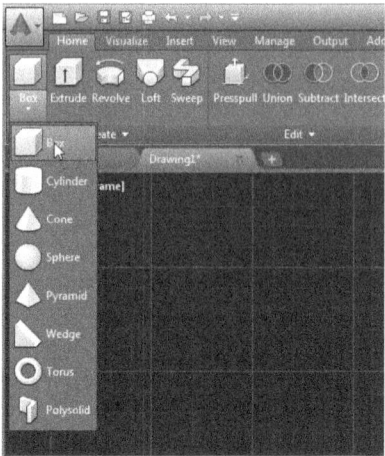

Pic 5.3 Click Box icon

2. Insert the first point, and insert the second point.

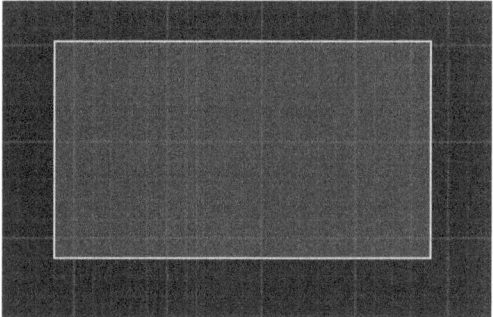

Pic 5.4 Create rectangle for box

3. Drag mouse to the top-right.

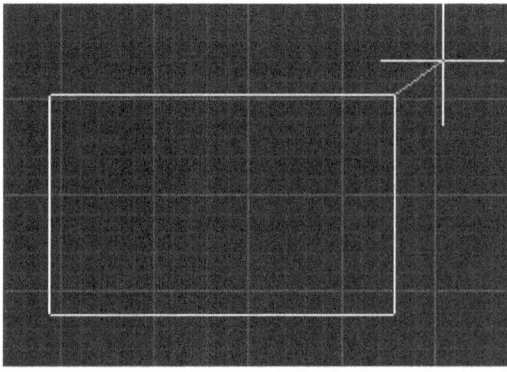

Pic 5.5 Drag mouse to top right

4. Insert the height of the box, for example 300.

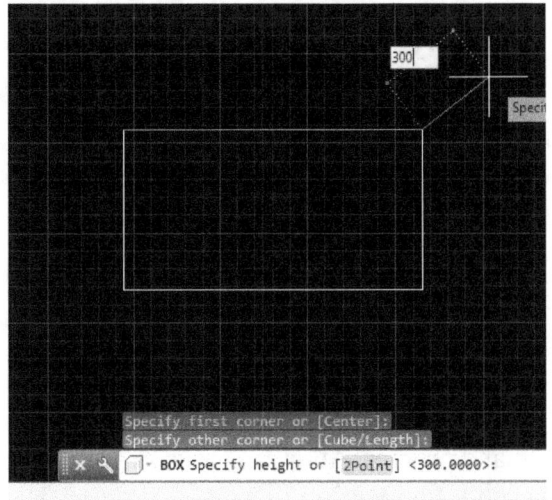

Pic 5.6 Specify the height of the box

5. To see the result in 3D, change the orbit button on the right toolbar.

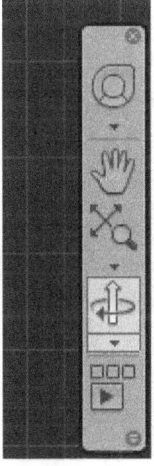

Pic 5.7 Changing the orbit

6. The result as below:

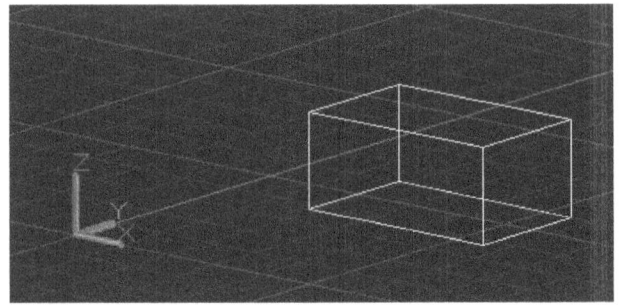

Pic 5.8 Result of box creation

7. Click Esc or [ENTER] in keyboard.

Other example:

1. Repeat step number 1-2.

2. In the command prompt, AutoCAD asks to specify other corder or length. Choose L for length.

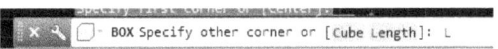

Pic 5.9 Choose L

3. It means, we'll draw by inserting length.

4. AutoCAD asks the length, type: 100.

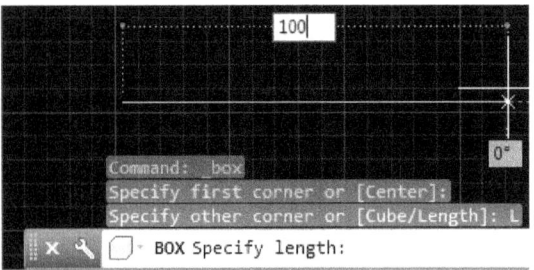

Pic 5.10 Displaying object's length

5. AutoCAD asks the width, type: 40.

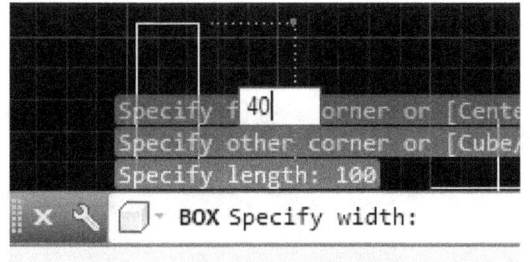

Pic 5.11 Specify width

6. Then drag top-right and specify the height to 50.

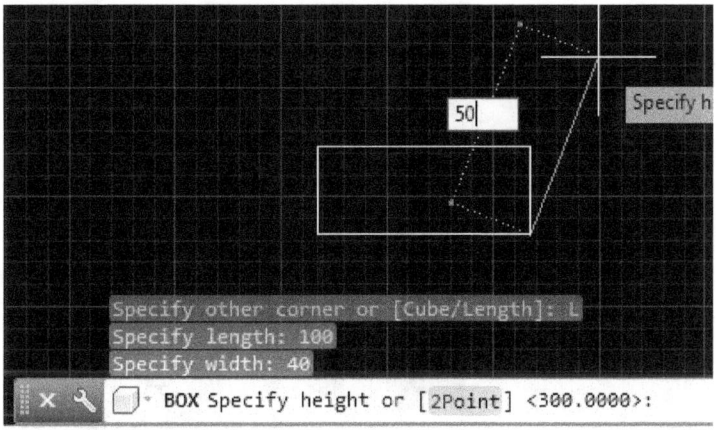

Pic 5.12 Specify height

7. The box result will be 100 x 40 x 50.

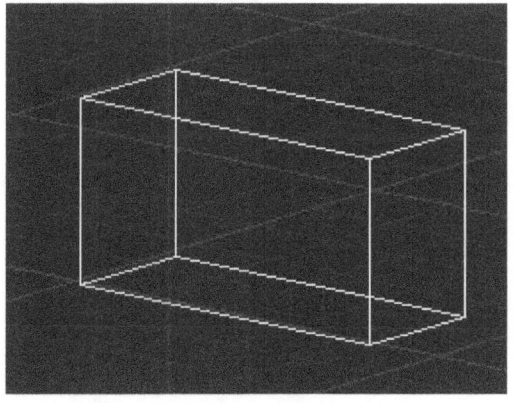

Pic 5.13 Box created

For drawing a cube, you can do these steps:

1. Execute Box, then type C for choosing cube.

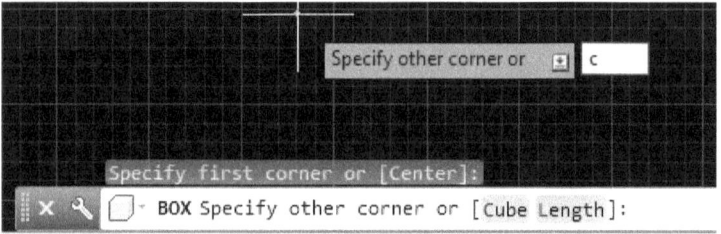

Pic 5.14 Specify C for cube

2. Specify the length to 100.

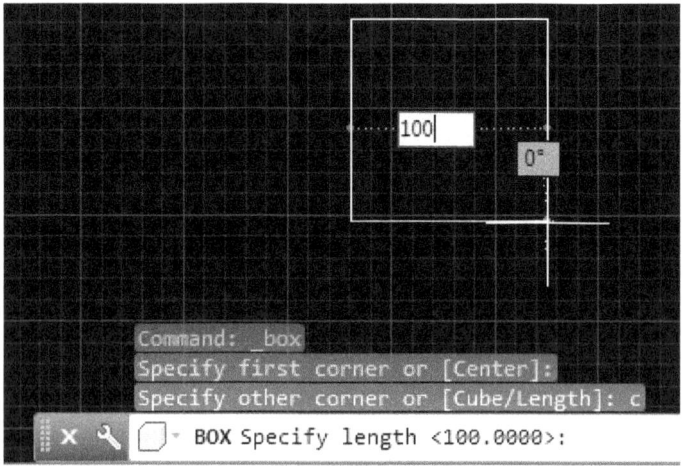

Pic 5.15 Specify the length of cube

3. The result is a cube with size: 100 x 100 x 100.

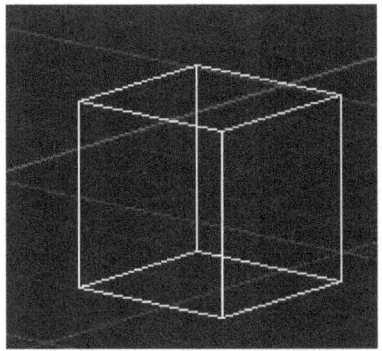

Pic 5.16 Cube already created

5.3.3 Draw Cylinder

The cylinder on AutoCAD created using Cylinder icon. See steps below:

1. Click on cylinder icon.

Pic 5.17 Click on Cylinder icon

2. Enter the center of cylinder. Click on a certain place.

3. Drag mouse to draw a circle.

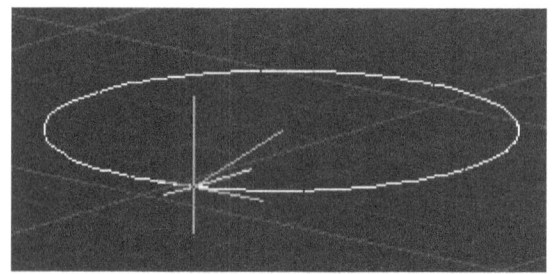

Pic 5.18 Drag mouse to draw a circle

4. Insert the height of the cylinder: 500.

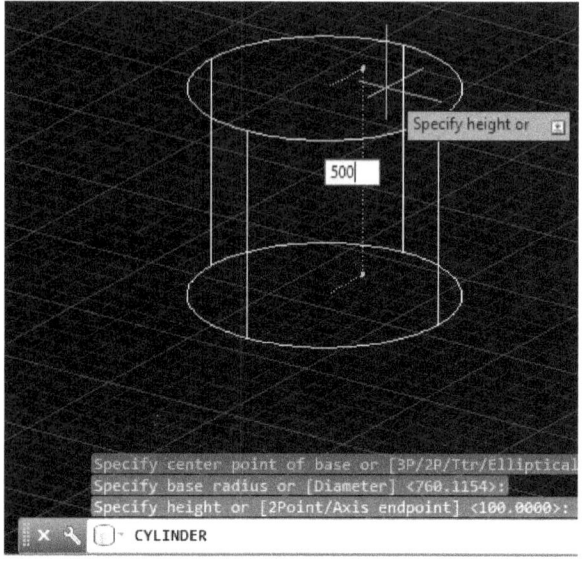

Pic 5.19 Inserting the height of cylinder

5. The result will be like this.

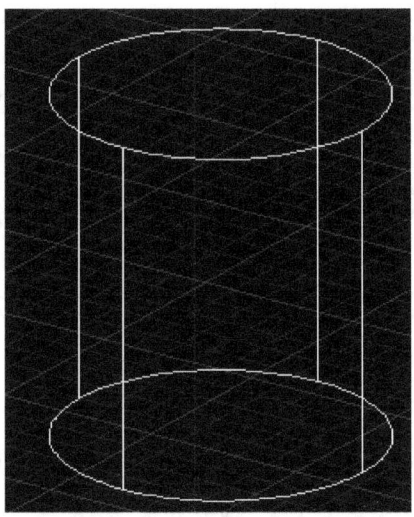

Pic 5.20 Cylinder already created

Another method is by defining radius or diameter. Follow steps below:

1. Repeat steps until step 2

2. Choose d for diameter.

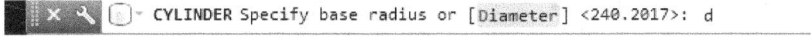

Pic 5.21 Choose D

3. This means, cylinder will be drawn based on diameter.

4. AutoCAD asks for diameter, type: 100.

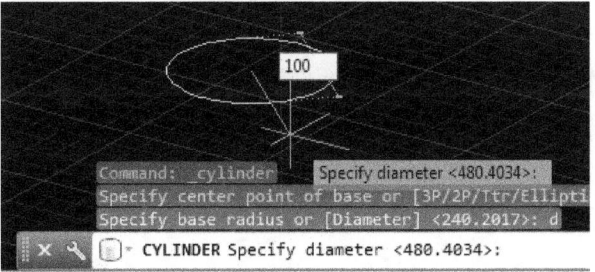

Pic 5.22 Specify the diameter

5. Specify the height: 200.

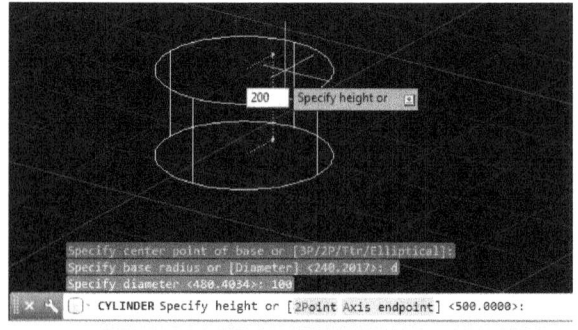

Pic 5.23 Specify the height

6. The result is cylinder with diameter =100 and height = 200

5.2.3 Draw Cone

Cone in AutoCAD can be created using Cone icon. See steps below to create Cone:

1. Click Cone icon.

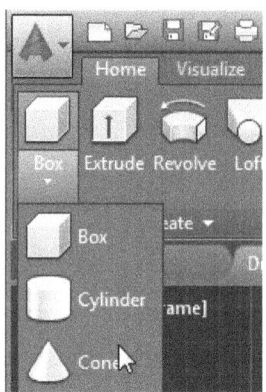

Pic 5.24 Click on Cone icon

2. AutoCAD asks to specify the center of the circle.

3. Drag mouse to draw a circle.

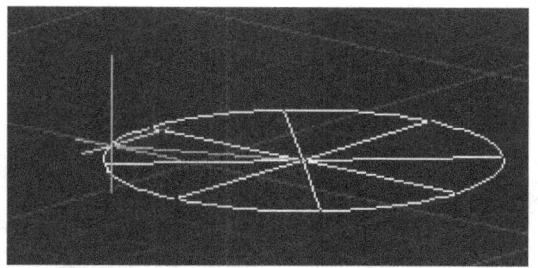

Pic 5.25 Drag mouse to draw a circle

4. AutoCAD asks for height for the cone.

5. The result will be like this.

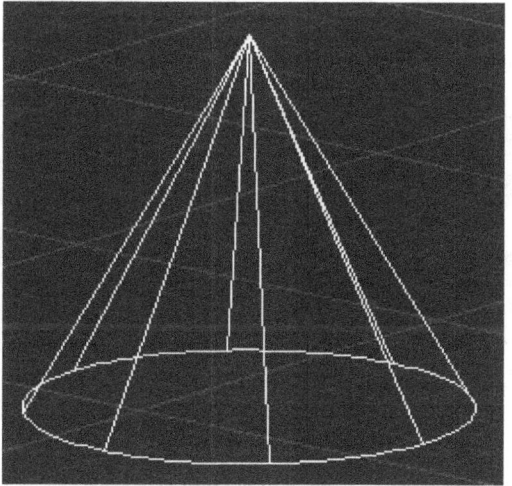

Pic 5.26 Cone result

Or you may specify the diameter and height by using steps below:

1. Repeat steps until step 2

2. In the command prompt, select D to insert diameter.

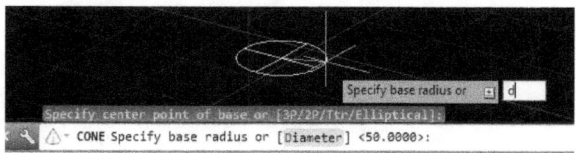

Pic 5.27 Select D

---

3. It means the circle will be created using diameter.

4. Insert the diameter, for example: 100.

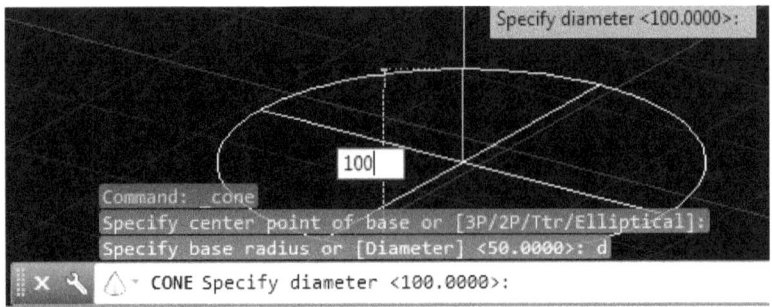

Pic 5.28 Inserting diameter for the circle

5. Specify the height for the cone, for example:150.

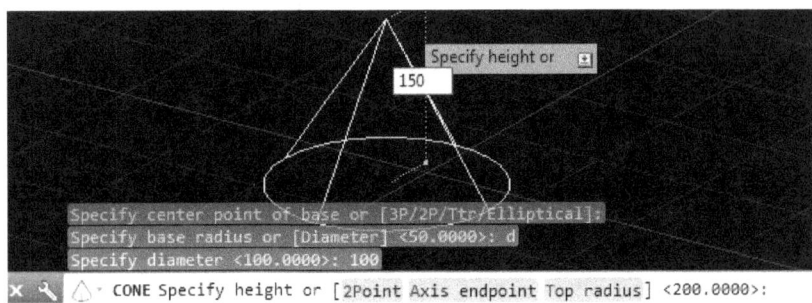

Pic 5.29 Specifying height of the cone

6. The result is cone with diameter = 100 and height = 150

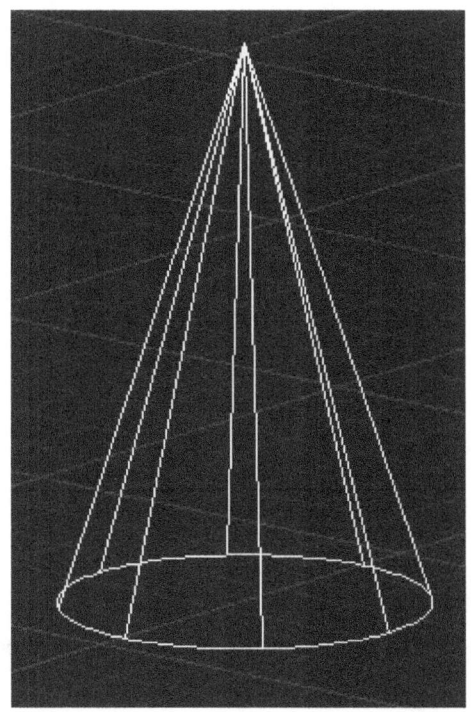

Pic 5.30 Cone result

5.2.4 Draw Ball

Ball can be created using sphere icon. See steps below for drawing ball in AutoCAD:

1. Click Sphere icon.

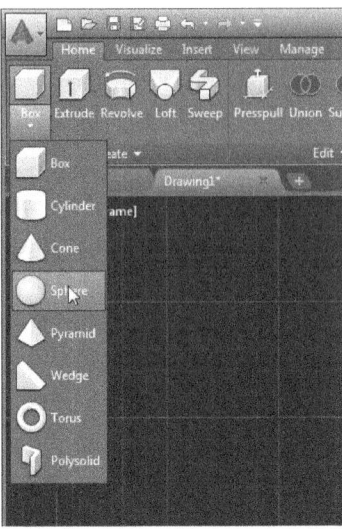

Pic 5.31 Click Sphere icon

2. AutoCAD asks the circle midpoint. Click the midpoint.

3, Drag mouse to draw the ball.

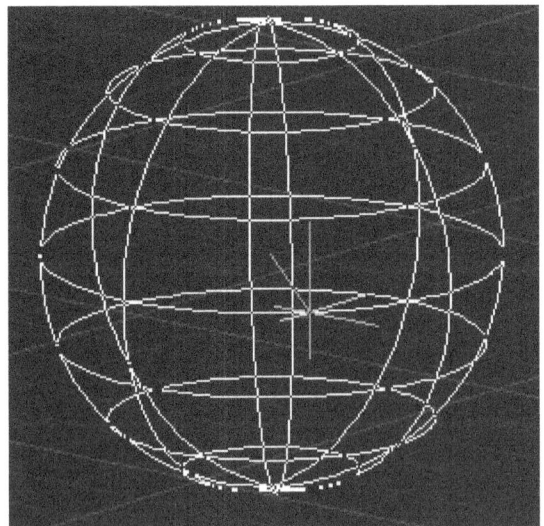

Pic 5.32 Ball result

4. You can see the result on the pictue above.

Another method is by specifying radius or diameter:

1. Repeat steps until step 2.

2. Insert d to specify diameter.

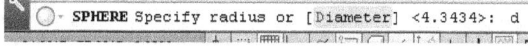

Pic 5.33 Choose Diameter

3. Insert the diameter, for example: 100.

5. The result is a ball in diameter = 100.

5.2.5 Draw Pyramid

You can draw pyramid using steps below:

1. Click pyramid icon,

Pic 5.34 Choose Pyramid

2. Specify the center of the rectangle.

3. Drag mouse to create the rectangle.

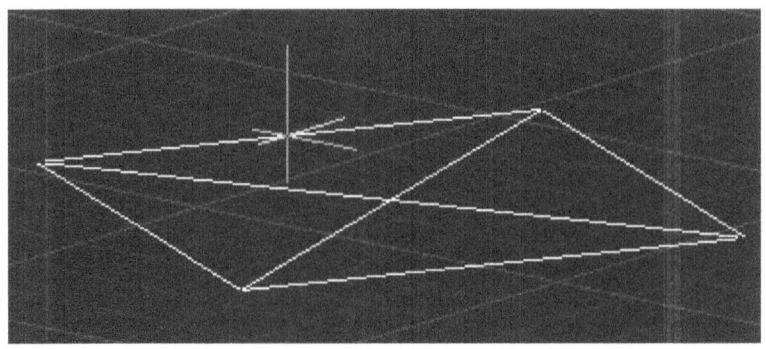

Pic 5.35 Create rectangle

4. Insert the height.

5. See the result in picture below.

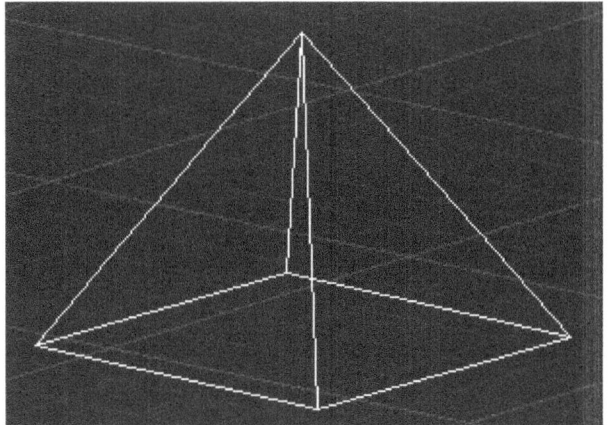

Pic 5.36 Pyramid result

5.2.6 Draw 3D Donut

You can also draw 3D donut using torus command. See example below:

1. Click on **Torus** menu.

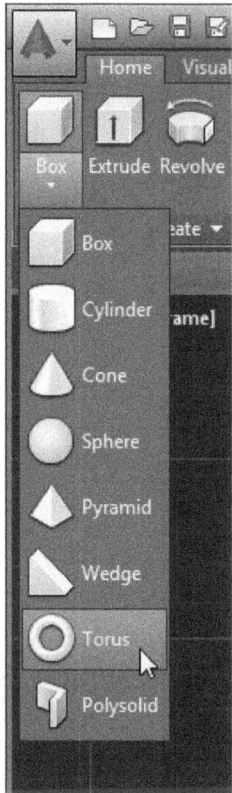

Pic 5.37 Click on Torus menu

2. Enter the center of the circle.

3. Drag the mouse.

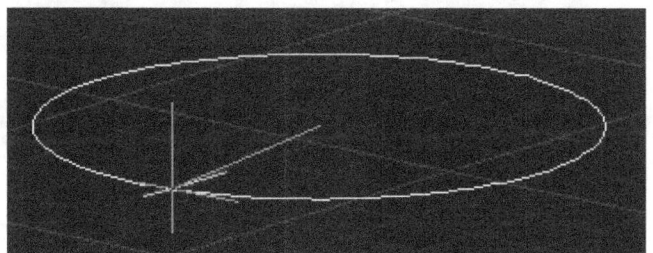

Pic 5.38 Drag the mouse from the center outwards

4. Insert the radius of the small circle of the tube.

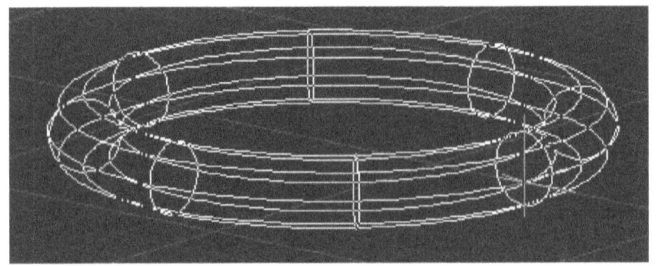

Pic 5.39 Creating the torus

5. Finish.

You can also make torus manually. See steps below:

1. Repeat steps above until step 2.

2. Choose radius, type: 50.

3. The result is a ring in radius = 50.

4. Then specify the radius of the tube = 10.

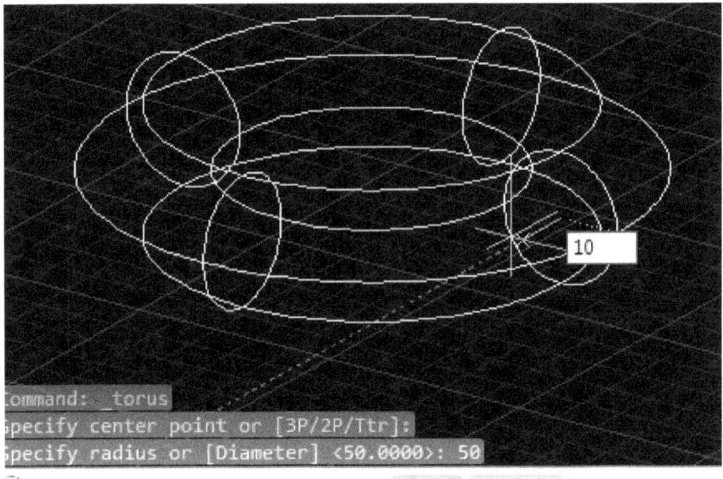

Pic 5.40 Specify the radius of the tube

5. The result is a torus with radius = 50 and tube radius = 10.

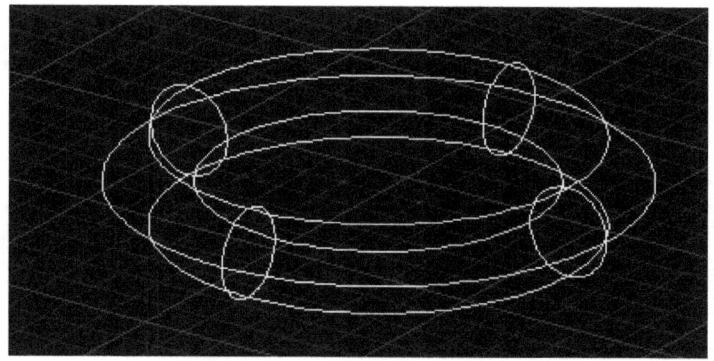

Pic 5.41 Result of ring drawing

5.2.7 Extrude 2D Object

You can draw a 3D object by extruding a 2D object. See example below:

1. Open the 2d pic.

2. Click orbit button.

3. Right click and drag to change the view of the 2d pci.

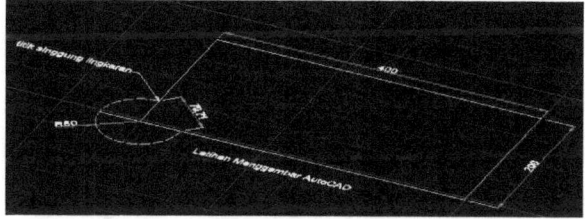

Pic 5.42 2D pic

4. Clik Extrude button like picture below:

Pic 5.43 Click Extrude button

5. Change the object youw ant to extrude, and click **Enter**.

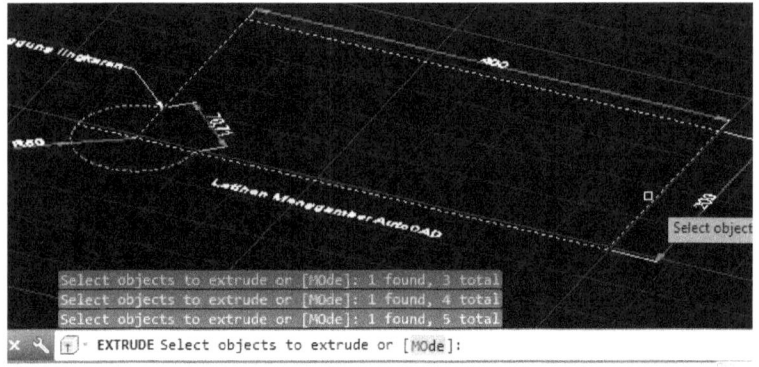

Pic 5.44 Change the object to extrude

6. Insert the height, for example: 1000.

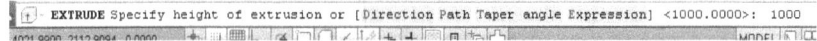

Pic 5.45 Insert the height

7. The 2d pic will be 3D right now, with height = 1000.

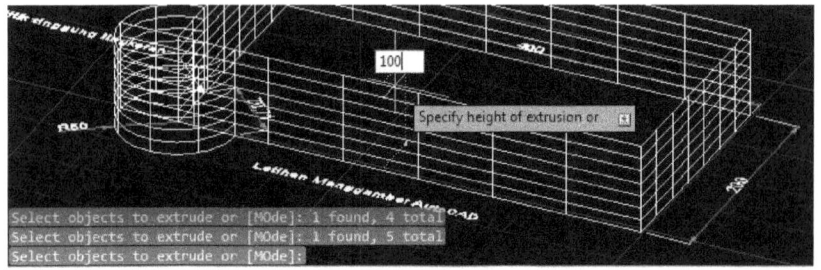

Pic 5.46 Result of extrude process

✓ *Exercise Extrude Feature*

You can also extrude your sketches with the Extrude feature. Draw a Polygon and set the number of edges to 8. Then chose CenterPoint as the center, and choose between Circumscribed or Inscribed within a circle. Finish the Polygon and type in "Extrude." Select the Polygon as a base. Type in "Mode" followed by "Solid" to create a solid 3D Object. Then set the height of the object. You can change the height by double-clicking the object.

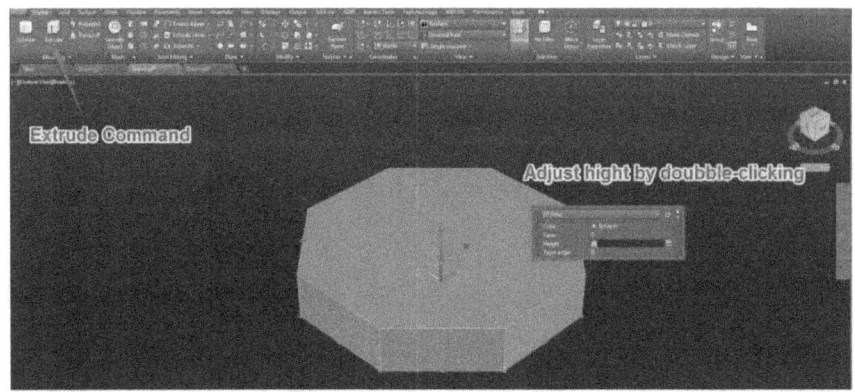

5.2.7 Chamfer and Fillet Feature

Edges and corners can be smoothed or chamfered easily. Switch to the Solid tab and click on Fillet Edge. You can now select all the top edges of the polygon. To lower the effort selecting all the edges manually, type in "Loop". Then click on one top edge. Click next to flip through the possible edge connections. When all the top edges are highlighted click accept and type "Radius" to define the size of the Fillet. You can try out different values and preview the fillet. Click or type in radius again to change it. Press enter two times to accept the previewed fillet.

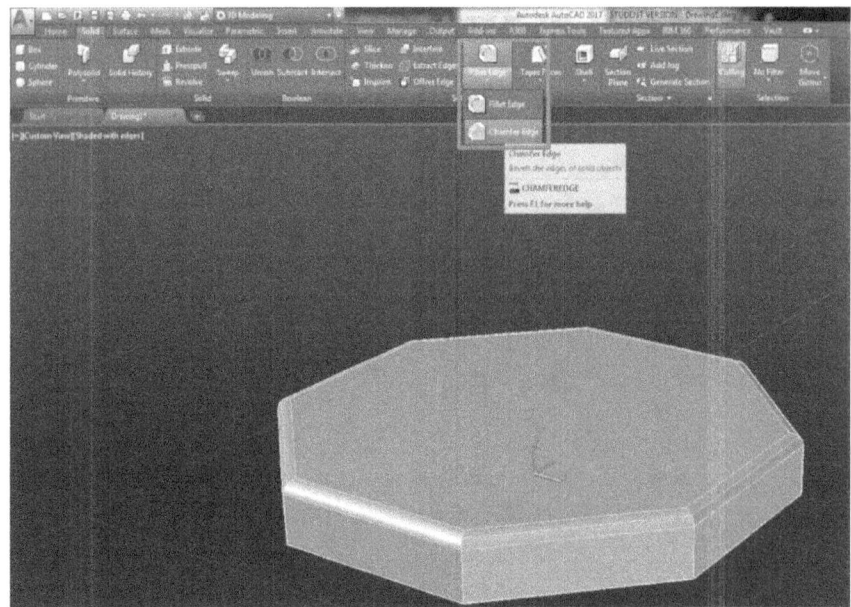

Pic 5.47 Chamfer and Fillet Feature

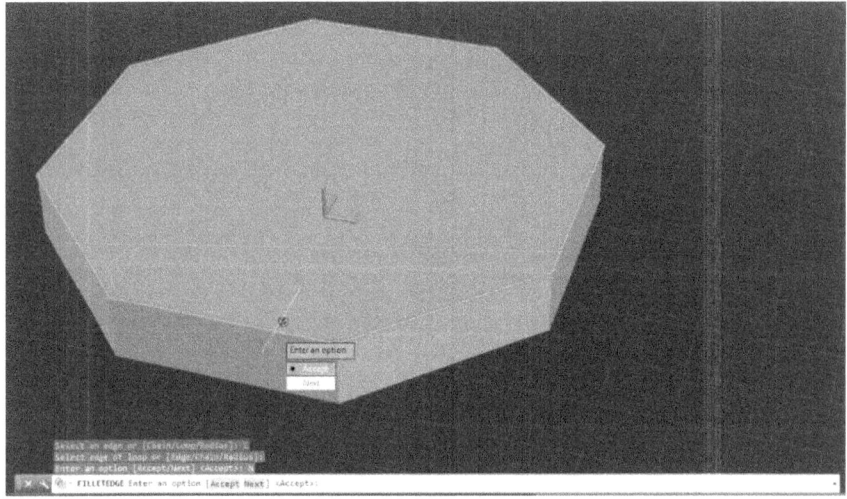

Pic 5.48 Fillet process

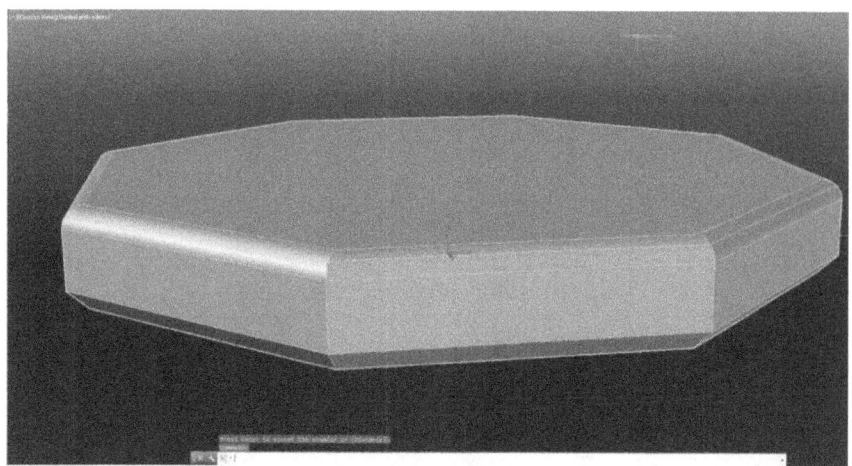
Pic 5.49 Result of fillet process

Now turn the polygon around and select with the arrow under the fillet feature the chamfer feature. Type in "Loop again and select one bottom edge of the polygon. Click Next until the lower rim of the polygon is highlighted followed by Accept. Now click on Distance and type in the first length of the chamfer. Confirm by pressing enter and type in the second length. Once again you can see a preview and hit enter two times to confirm.

5.2.8 Merge, Subtract, and Intersect 3D Objects

Build a sphere with the same radius right on top of a cylinder. Now type in "Union" and select the sphere and the cylinder. Confirm with Enter. When you hover over both shapes you will see that they have become one solid object.

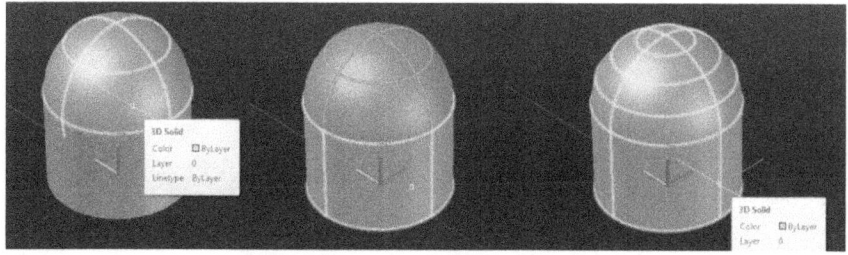
Pic 5.49 Result of merge process

Repeat build a cylinder and sphere or use Undo to the point before you merged both objects. Now type in "Subtract." At first, you need to select the object which to subtract from. Select the cylinder and confirm. Now select the sphere as the object to subtract and confirm.

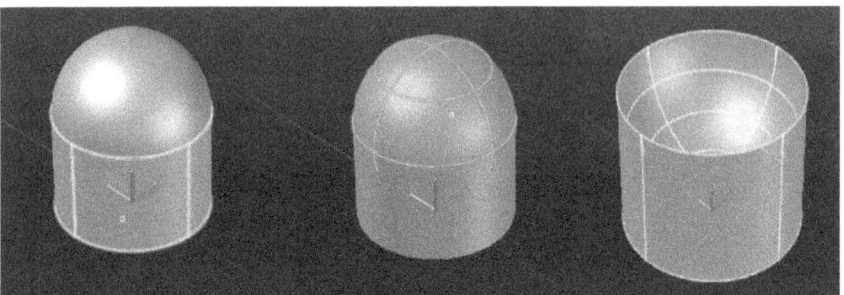

Pic 5.50 Result of subtract process

Start with the single sphere and cylinder again. Now type in "Intersect," select both objects and confirm.

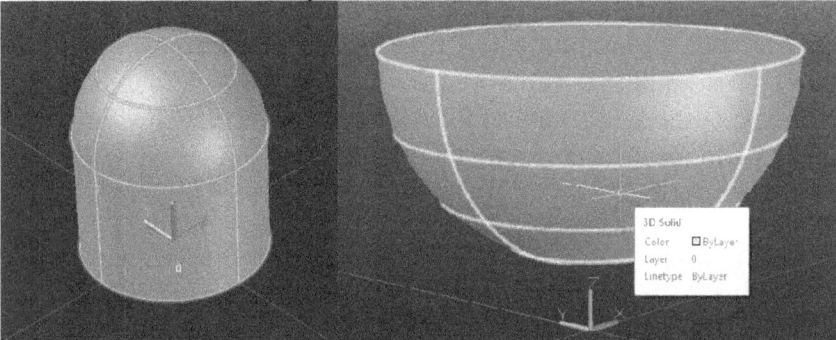

Pic 5.51 Result of intersect process

CHAPTER 6 MESH-FILES IN AUTOCAD

If you want to 3D print or share your creations with other people, you probably want to create or edit a mesh file like .stl. However, AutoCAD is not the ideal CAD software for this matter. It can export to .stl, but it sadly can not open .stl or .obj files. There are however ways to bypass this problem.

6.1 Import .stl and other Mesh-Files

As stated, AutoCAD cannot import Mesh-Files, but it can work with the standardized ISO format STEP .step and Autodesk's interchange format .dxf. To generate these file types you can use other AutoCAD software like Inventor or free software like FreeCAD. You can also use a quick way and upload the .stl to a converter provided by CAD-Forum and generate a .dxf file.

Open .dxf files in AutoCAD by first creating a new Drawing. Then click on the AutoCAD logo > Open > Drawing and select .dxf as file type in the file browser. When the model is imported, you can change the visual style by typing VISUALSTYLE.

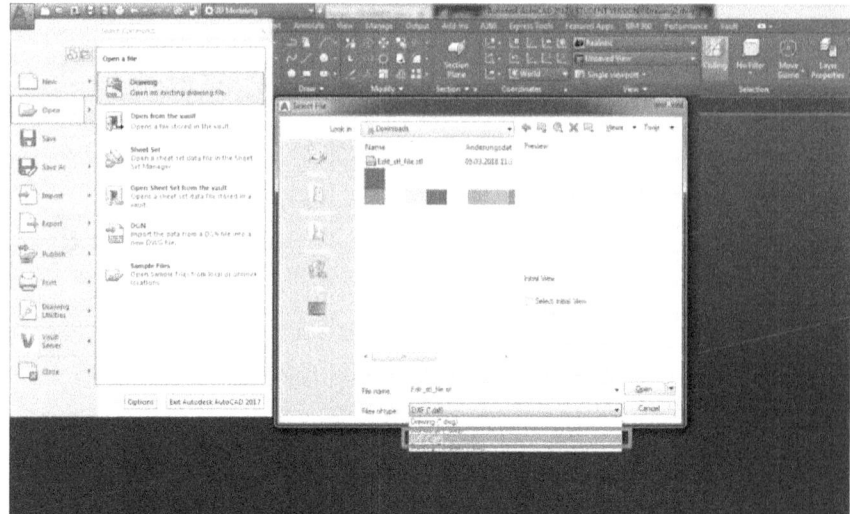

Pic 6.1 Import mesh file process

Pic 6.2 Result of import mesh file process

6.2 Export .stl

Fortunately, exporting .stl files is possible with AutoCAD. Click on the AutoCAD logo > Export > Other file formats and select .stl as the file type in the file browser.

CHAPTER 7 CREATE TECHNICAL DRAWING

If you want to create a technical drawing of y model you created, AutoCAD is a great software to work with. At first, you will need a template sheet for the technical drawing. You can find templates on the AutoCAD website for free. Download the Manufacturing Metric template. Open the object you want to create a technical drawing from. Then right-click on the + in the bottom left corner and open the downloaded template. You can insert your name, project or other information into the title block in the bottom right of the sheet by double-clicking it.

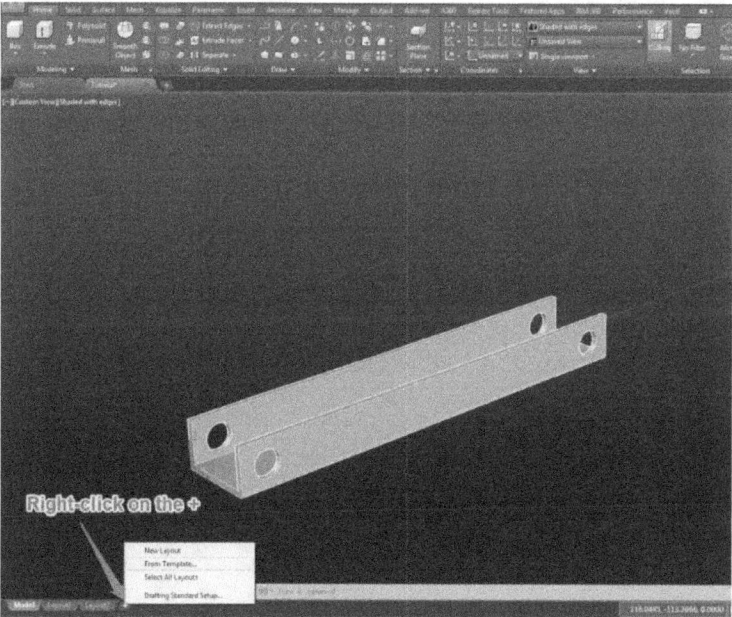

Pic 7.1 Result of import mesh file process

7.1 Insert Model Views

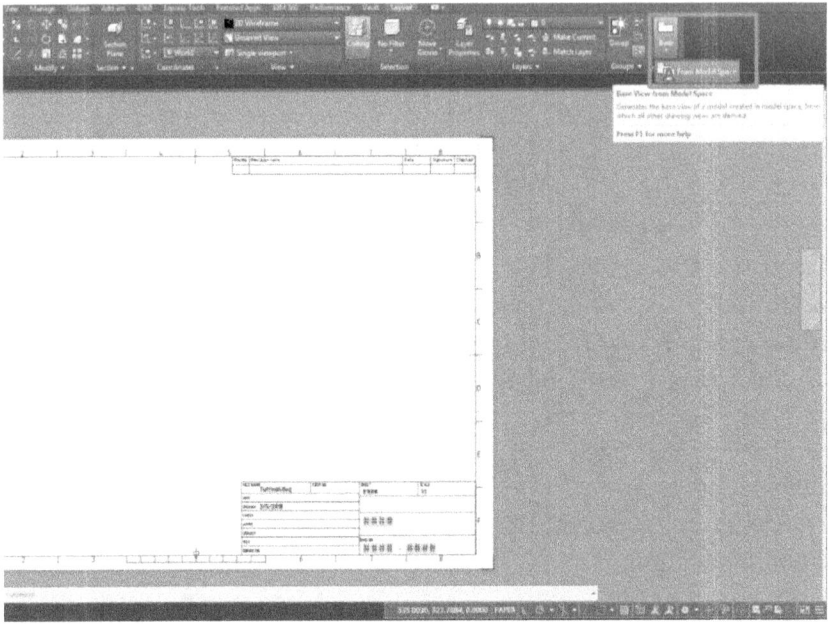

Pic 7.2 Insert Model Views

Once you are in the drawing sheet template tab, click on Base > From Model Space.

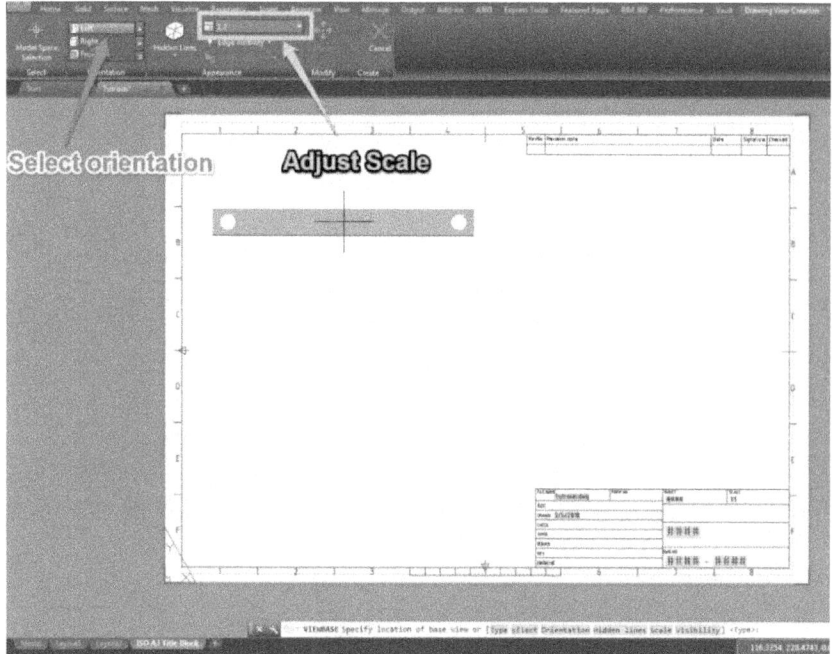

Pic 7.3 Select orientation to switch to a different view

Click to place the first view (which is the front view) in the middle of your sheet. Once you have clicked, you can select orientation to switch to a different view. If the model is too large or small, click on Scale and select a scaling factor. Click on Move to position the object. Left-Click at the desired position to accept. You can now continue to place other views by dragging the mouse horizontally or vertically. Left-Click to confirm each position. If you move the object to a 45° angle, you can place the isometric view. Try to place enough views of the object so most or all of the features can be seen. If you select one view, you can move it with the blue square and size it with the blue triangle.

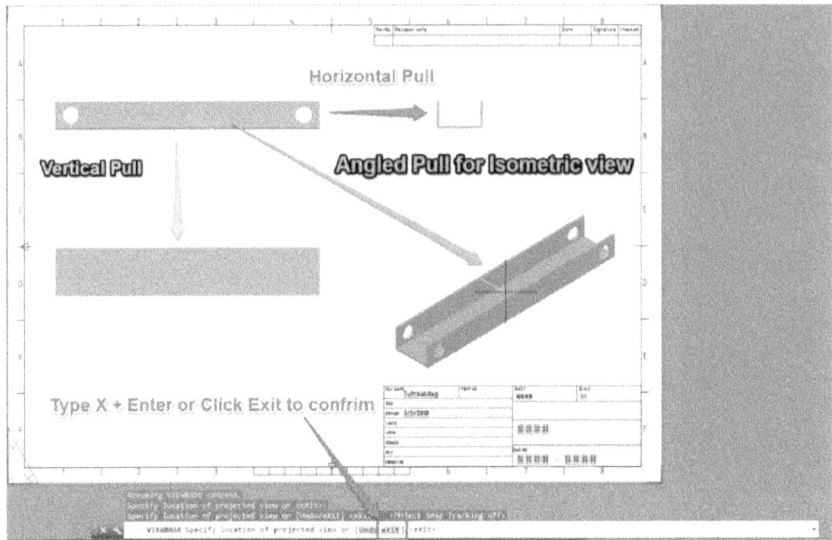

Pic 7.4 Result of Model Views

7.2 Place Dimensions

When placing dimensions, you have to follow three basic rules:

1. start with the smallest detail

2. Annotate a detail only once

3. Annotate every detail

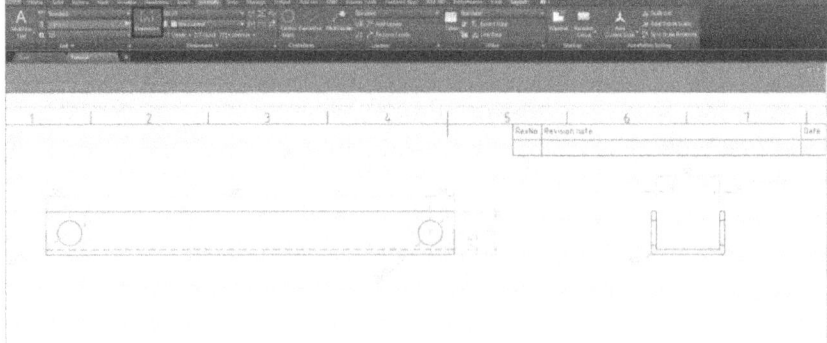

Pic 7.5 Dimension of Objects

To start annotating switch to the Annotation tab. Select the Dimension command. This is a smart command which adapts to the feature you want to annotate. Now select the first line or two dots you want to describe. You will then see the length or radius, and you can move the annotation into position. Position the annotation, so it does not intercept with other lines, numbers or is too close to the object itself.

If you want to dimension circles or holes, you will have to place a center mark first. Click on Center Mark in the annotation tab and select a circle. Now use the Dimension Command to annotate the circle. You can switch between Radius and Diameter by typing R or D on your keyboard.

7.3 Detail and Section View

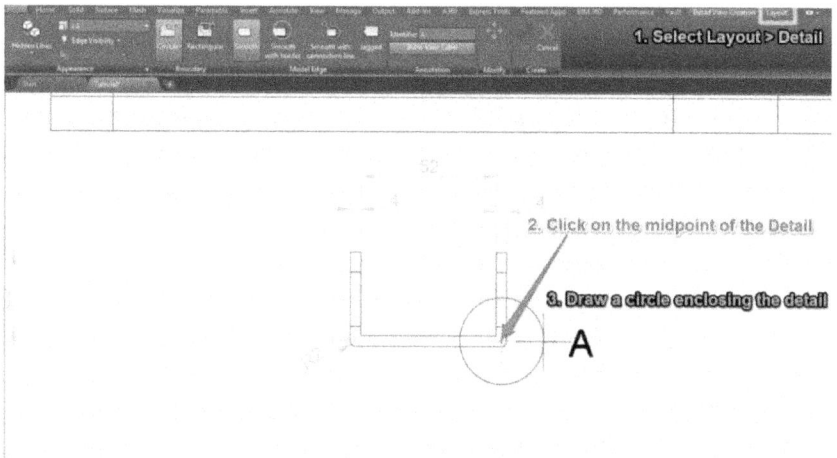

Pic 7.6 Select orientation to switch to a different view

To place a detail view of your drawing click on Layout > Detail > Circular. First, select the parent view you want to specify followed by clicking in the middle of the detail to set a center point. Then draw a circle enclosing the detail. Place the detailed view at a free spot.

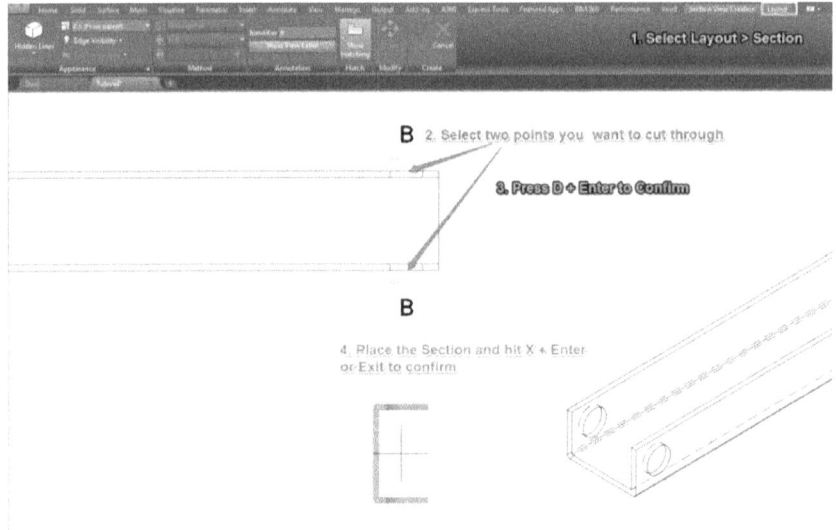

Pic 7.7 Select two points

If you want to look inside a drawing, you can use Layout > Section View. Select the view you want to create a section from followed by selecting two points for the section line. Confirm by pressing enter and place the section view at a free spot. You can also change the size and line style afterward.

This brings us to the end of our AutoCAD tutorial for beginners.

AutoCAD is a powerful CAD software, which is supposed to be used for architectural design and mechanical engineering. It has one of the best toolboxes and features to support 2D drawings. When it comes to 3D design, it is still impressive, especially when rendering 3D objects in a realistic way.

However, there are easier to use 3D programs. One major disadvantage of AutoCAD is the missing support of mesh files. You cannot import or export .stl or .obj when working with AutoCAD without going through some workarounds. There are some plugins, yet they only support binary mesh files. Still, AutoDesk offers another 3D software called Inventor, which is great for creating or editing 3D models. You can access it with your student license or use it with your 3-month free trial.

***

ABOUT THE AUTHOR

Zico P. Putra is a senior engineering technician, CAD consultant, author, & trainer with 10 years of experience in several design fields. He continues his PhD in Queen Mary University of London. Ali Akbar is an AutoCAD Author who has more than 10 years of experience in the architecture and has been using AutoCAD for more than 15 years. He has worked on design projects ranging from department store to transportation systems to the Semarang project. He is the all–time bestselling AutoCAD author and was cited as favorite CAD author. Find out more at https://www.amazon.com/Zico-Pratama-Putra/e/B06XDRTM1G/

CAN I ASK A FAVOUR?

If you enjoyed this book, found it useful or otherwise then I would really appreciate it if you would post a short review on Amazon. I do read all the reviews personally so that I can continually write what people are wanting.

If you would like to leave a review, then please visit the link below:

https://www.amazon.com/dp/B06XS99PKP

Thanks for your support!

www.ingramcontent.com/pod-product-compliance
Lightning Source LLC
Chambersburg PA
CBHW072135170526
45158CB00004BA/1387